# WALKER'S QUANTITY SURVEYING AND BASIC CONSTRUCTION ESTIMATING



An Introductory Book Explaining
Construction Estimating

Written By

**G. PATRICK BOURGEOIS**

**LUTHER J. STRANGE**

**BRISBANE H. BROWN, JR.**

**C. DAWSON ZEIGLER**

School of Building Construction
University of Florida
Gainesville, Florida

**FRANK R. WALKER COMPANY**

Publishers
Lisle, Ill.

Every attempt has been made to make this book as accurate and authoritative as possible. We will be grateful to readers of this volume who will kindly call attention to any errors, typographical or otherwise. We also invite constructive criticism and suggestions that will make future editions of this book more complete and useful.

Neither the Frank R. Walker Co., the editor, the author nor the sources listed in the back of this book guarantees or warrants the correctness or sufficiency of the information contained in this book; nor do any parties assume any responsibility or liability in connection with the use of this book or of any products, processes or pricing described in this book.

FRANK R. WALKER COMPANY



Publishers of
Walker's Building Estimator's Reference Book
Walker's Practical Accounting and Cost Keeping for Contractors
Walker's Insulator's Estimating Handbook
Walker's Manual For Construction Cost Estimating
Walker's Remodeling Estimator's Handbook
Bookkeeping Systems and Forms For Contractors



**ISBN: 0-911592-75-X**

Printed in the U.S.A.

## FOREWORD

Estimating is one of the most important skills needed by today's building construction contractors. Educating an estimator is a long and formidable task. You start with an intelligent, highly motivated individual who understands the construction process and educate him or her in how to read construction drawings and specifications, construction materials, techniques, structures and productivity, quantity survey and estimating, and add several years of experience working for a qualified estimator and you have an estimator.

This book is intended to be used in conjunction with *Walker's Building Estimator's Reference Book* by Frank R. Walker for a study of building construction quantity surveying, and basic estimating from the viewpoint of the general contractor. The organization and format can be easily adapted by specialty contractors.

The concept and organization of this book are the efforts of C. Dawson Zeigler who has worked many years as an estimator and who teaches estimating. Luther Strange supplied some of the exercises and he worked closely with G. Patrick Bourgeois who wrote the book and drew the figures.

Many people had recognized the need to reduce this material to a book and my only contribution was to bring the resources together.

Brisbane H. Brown, Jr., P.E.
Professor and Director
School of Building Construction
University of Florida
Gainesville, FL 32611

# PREFACE

This text is primarily intended for readers with prior background in construction techniques and construction contracts. The author's intent is to guide the readers through a proven method of quantity surveying and estimating. The method was developed by the School of Building Construction at the University of Florida to provide a standardized method of quantity surveying. This standardized method allows several estimators to work on the same project without confusion or loss of time. It also allows one estimator to take over an unfinished estimate for an absent estimator and understand what has been completed.

Part One of this text will be a review of construction contracts, insurance and bonds. This section will also contain a review of geometric shapes, conversion factors and tables which will aid the reader in determining material quantities. The preparation of the plans for the survey will be discussed including how to read plans and locate sections, and the conversion of dimension from inches to decimals of a foot.

The second part will explain the logical approach method to quantity surveying. The format for the quantity survey will be introduced. This section will provide the reader with a method of estimating quantities and cost of concrete, formwork, reinforcing steel, masonry, earth work and carpentry.

The final section will deal with subcontractors and how to incorporate subcontractor's bids into your estimate. A method for estimating insurance, bonds, taxes, and workman's compensation will be developed. The final chapter will show how to transfer all the various cost items to a summary sheet. The summary sheet is an aggregation of all items of cost, developing a total cost of a structure.

This text provides a guide for residential estimating. After completion of this manual, the reader should be able to prepare a complete cost estimate for a single-family residence or a small commercial job. The manual can be used as a preliminary guide to commercial estimating with further studies in commercial techniques.

G. Patrick Bourgeois

Gainesville, Florida

# TABLE OF CONTENTS

# LIST OF TABLES AND FIGURES

## TABLES

PAGE



## FIGURES

**FIGURES** **PAGE**

CHAPTER 1

# CONSTRUCTION REVIEW

## Construction Contracts

The essential documents usually found in a construction contract are as follows:

1. General Conditions
2. Supplementary Conditions
3. Technical Specifications
4. Drawings
5. Addenda
6. Agreement

Collectively, all the foregoing documents constitute the construction contract and must be studied so that an accurate cost estimate can be prepared.

The general conditions and supplementary conditions are considered job overhead. These are items the general contractor must furnish to conduct the job as a whole. Each job will entail expenses that are chargeable to that specific job. These will include the superintendent, temporary building, enclosures, barricades, offices, toilets, utilities, protection, clean-up, permits, surveys, photographs, tools and equipment, insurance, sales tax, surety bonds and warranties.[1] The quantities and cost of these items must be estimated and included in the total cost estimate of the structure. As a group these are called "Indirect Costs".

The technical specifications prescribe the materials and the workmanship standards required. Generally speaking, the drawing shows what is to be built, and the specifications describe how the project is to be constructed and what

[1]Walker, Frank R., *The Building Estimator's Reference Book,* (Frank R. Walker Company, Chicago, Ill.)

results are to be achieved. The specifications provide information on the type of concrete or mortar required, and the brand name of doors, windows, hardware, carpeting, etc. These items are usually prescribed by the architect and the cost for these items must be included in the cost estimate.

The drawings or plans portray the physical aspects of the structure, showing the arrangement, dimensions, construction details, materials and other information necessary for building the project. The drawings for a building typically include the plot plan, structural, architectural, plumbing, mechanical and electrical classifications. In most cases, the general contractor is responsible for estimating the quantities and cost of items in the plot plan, structural and architectural drawings. The general contractor may choose to subcontract items like brick, masonry, steel, roofing and carpentry to a specific contractor. In most cases the general contractor will subcontract the plumbing, mechanical and electrical classifications. In any case, the estimator should be familiar with all the drawings to check the subcontractor's bid.

The addenda are used to notify the estimator of changes, corrections, clarifications or interpretations of the contract documents. The addenda should be numbered so that the estimator can make sure he has checked all the addenda for cost modifications. The AIA General Conditions require that such corrections, changes, interpretations and clarifying answers on the part of the architect-engineer be in writing. It is the responsibility of the architect-engineer to see that copies of all addenda promptly reach the parties who receive plans and specifications, but it is the responsibility of the estimator to include the addenda in the cost estimate.

The agreement is a document specifically designed to formalize the construction contract. Although there are many different types of construction contracts, they may be grouped together into two large divisions. One division includes those contracts for which the contractor is selected on the basis of competitive bidding. Competitive bid contracts are customarily prepared on a fixed price basis requiring the estimator to come up with a lump-sum amount for the structure. The second major division of construction contracts consist of those that result from direct owner-contractor negotiation. A negotiated contract can be one of many different forms; lump-sum, unit price, or cost plus fee. Most negotiated contracts are on a cost-plus-fee basis in which the owner reimburses the contractor for all construction cost and compensates him for his services. No matter what type of agreement between the contractor and the owner, the estimator will be required to make a survey of materials required for the structure.

### Construction Surety Bonds

A surety is a party that assumes legal liability for the debt, default, or failure in duty of another. In construction, bonds are typically used to protect owners, workers and material suppliers. If the contractor defaults the surety becomes liable to these parties but the contractor is still liable to the surety. The estimator can determine what bonds are required by studying the general conditions of the contract. Some of the types of bonds encountered are discussed below.

Bid bonds are used to insure that bids will be submitted only by qualified contractors who will not only enter into the contract, but qualify for performance and payment bonds. The bid bonds guarantee that the contractor, if low bidder, will sign the contract and furnish the required performance and payment bonds. Typically the general conditions require the contractor to furnish a bid bond for five percent of his bid. If the contractor does not sign the contract, the surety company will be required to pay the difference between the low bid and that of the next acceptable bidder, but not exceeding the amount of the required bid bond. Surety companies usually require contractors to qualify for performance and payment bonds before they will issue a bid bond.[2]

Peformance bonds guarantee that the contract will be performed and that the owner will receive his structure. This bond customarily covers any warranty period that may be required by the contract. "All performance bonds have a face value which acts as an upper limit of expense that the surety will incur in finishing the contract should that action become necessary."[3] This face value is specified in the general conditions and is usually derived from some percentage, 100 percent, for instance, of total contract price.

Payment bonds act primarily for the protection of third parties to the contract and guarantee payment for labor and materials used or supplied in the performance of the construction. The private owner is protected against liens that can be filed on this project by unpaid parties. All government works projects require payment bonds to protect labor and material suppliers, because liens cannot be filed against the government.

The general conditions may require other bonds, but the bid bonds, performance bonds and payment bonds are the most common. Performance and payment bonds cost depends on the amount of the contract and the length of coverage. "Bonds will usually cost $10.00 per $1,000.00 of contract cost for the first $100,000.00; $6.50 per $1,000.00 for the next $2,400,000.00; $5.25 per $1,000.00 for the next $2,500,000.00; $5.00 per $1,000.00 for the next 2,500,000.00 and $4.70 per $1,000.00 over $7,500,000.00."[4] The cost of bonds must be established by the estimator and included in the cost of the project.

[2]Walker, Frank R., *The Building Estimator's Reference Book,* (Frank R. Walker Company, Chicago, Ill.)

[3]Clough, Richard, H., *Construction Contracting,* (John Wiley and Sons, New York, 1977), p. 149.

[4]Walker, Frank R., *The Building Estimator's Reference Book,* (Frank R. Walker Co., Chicago, Ill.)

### Construction Insurance

Construction work is extremely hazardous, and accidents are frequent and often severe. The potential severity of accidents and frequency with which they occur require that the contractor protect himself with a variety of complex and expensive insurance coverages. The general conditions typically make the contractor responsible for obtaining coverage, such as Workmen's Compensation Insurance, Contractor's Public Liability and Property Damage Insurance, and Contractor's Contingent Liability Insurance. Property insurance to protect the project and liability insurance to protect the owner may be made the responsibility of either the owner or the contractor, depending on the contract. The estimator must establish what insurance is required for the job and determine the cost related to the specific job. An excellent insurance checklist that will aid the estimator in determining what insurance is necessary for the job is shown in *The Building Estimator's Reference Book* by the Frank R. Walker Company.

### Reading Construction Drawings

Working drawings for each job are arranged in logical sequence and numbered to form a set. The set for a small job, like a house, usually has the following sequence:

1. Plot Plan
2. Foundation Plan
3. Floor Plan
4. Elevations
5. Details
6. Interior Elevations
7. Mechanical Equipment
8. Electrical

Sets of working drawings for larger buildings usually are numbered in a similar sequence. The architectural drawings are labeled A-1, A-2, A-3, etc.; the structural drawings are labeled S-1, S-2, S-3, etc.; the mechanical drawings are labeled M-1, M-2, M-3, etc.

Details and cross sections are usually referenced by their location on the page and the number of the pages. The symbol 4/A-3 tells the estimator that the section is the fourth drawing on page A-3. (Figure 1)

Some sets of working drawings are more detailed and thorough than others. This requires the estimator to use his judgment when interpreting the plans. He should use stated dimensions for all calculations except when no dimensions are given, then the drawings may be scaled. The estimator must take caution when scaling drawings and check other dimensions. Before the survey begins, he should familiarize himself with the plans checking all sections.

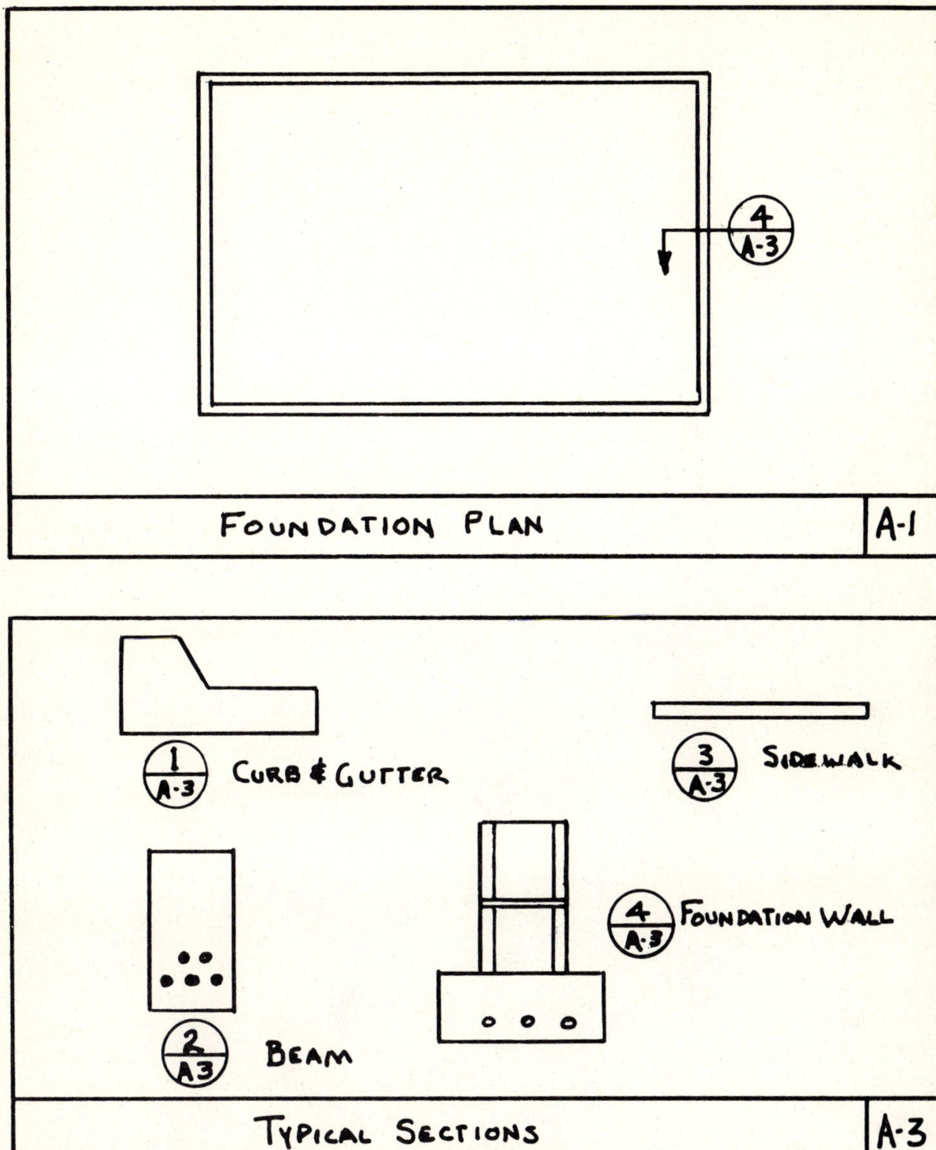
4
A-3
FOUNDATION PLAN
A-1
1
A-3
CURB & GUTTER
3
A-3
SIDEWALK
4
A-3
FOUNDATION WALL
2
A3
BEAM
TYPICAL SECTIONS
A-3

**FIGURE 1**

CHAPTER 2

# COMPUTING AREAS AND VOLUMES

In estimating, the process of examining the construction documents to determine the specific amount of material, supplies and equipment required to construct a project is called quantity surveying. Determining the amount of materials required for a project requires an understanding of geometric shapes and mathematics. The information in this chapter will provide the student with the required skills for a quantity survey.

In quantity surveying, all dimensions and quantities are converted from inches to tenths of a foot. By using decimals, it is possible to state the fractional parts of inches, feet and yards, in a smaller space than if regular fractions are used. The decimal equivalents of various numbers of inches are given in the following table:

**Conversion Table**
**Inches to Feet**

| | | | |
|---|---|---|---|
| ½" | .0417 | 6½" | .5417 |
| 1" | .0833 | 7" | .5833 |
| 1½" | .1250 | 7½" | .6250 |
| 2" | .1667 | 8" | .6667 |
| 2½" | .2083 | 8½" | .7083 |
| 3" | .2500 | 9" | .7500 |
| 3½" | .2917 | 9½" | .7917 |
| 4" | .3333 | 10" | .8333 |
| 4½" | .3750 | 10½" | .8750 |
| 5" | .4167 | 11" | .9167 |
| 5½" | .4583 | 11½" | .9583 |
| 6" | .5000 | 12" | 1.0000 |

**TABLE 1**

When surveying the quantities of most construction materials, the preceding table will provide the required accuracy. Surveying specialty materials like stainless steel, brass, copper, etc. may require converting fractions of an inch into decimals of a foot.

For example: 1 1/2" = 1/12 + 1 2/12 = 0.125 Feet

"Most of the estimator's computations involve measurements of surface and cubical contents and are stated in lineal feet (Lin.Ft.), Square Feet (Sq.Ft.), Square Yards (Sq.Yds.), Squares (SQS.) containing 100 Sq. Ft., Cubic Feet (Cu.Ft.), and Cubic Yards (Cu.Yd.)."[5]

The estimator may be required to determine the weights of building materials. Rebar, for example, is sold by the ton. Table 3 on page 24 shows the weights of standard sizes of reinforcing steel.

[5]Walker, Frank R., *The Building Estimator's Reference Book,* (Frank R. Walker Co., Chicago, Ill.)

CHAPTER 3

# ACCURACY AND THE DETAIL REQUIRED IN ESTIMATING

When preparing detailed estimates, the estimator should make sure he has not made errors in calculations. The method and format used in this manual will aid the estimator in checking for errors in calculations. This check for gross errors in calculations is discussed when the format is introduced.

The detail required depends on the type of job. Commercial work with massive quantities requires more detail than residential work. In commercial work, adjustments are required so quantities are not counted twice. For example, the building line must be adjusted to the center line of an eight foot wide commercial footing. By using the center line of the footing, the corners are not counted twice. A method for this adjustment is discussed later in this manual. In a sixteen inch wide residential footing, the adjustment is not required because the quantities in the corners are insignificant. Typically, the estimator should not go to such detail that the cost of estimating exceeds the cost of the item. As a general rule, if the building line is 6" or less away from the center line, use the building line. Like all rules, this one has exceptions, as we shall see in the Masonry Division (Division 4).

CHAPTER 4

# LOGICAL APPROACH TO ESTIMATING

The estimator should prepare his estimate in a logical order. This order should follow the order of construction activities from start to finish. This logical construction order allows the estimate to be used to check unit cost, establish crew size requirements and control job cost. The Construction Specifications Institute has developed a standardized order for the arrangement of specifications. This arrangement closely follows the logical sequence of construction activities and is called the Uniform Construction Index (UCI). The UCI is made of sixteen divisions which can be used by the estimator as a guide to a logical order. Appendix A shows a sample coding system based on the UCI order and numbering system heareafter referred to as the UCI order.

By using the UCI, the estimator has begun to develop a standardized method. The next step is to develop a standardized format which is understood and followed by all personnel in the company. The following format is easy to follow and provides a place for all important information.

The format uses a Quantity Sheet available at stores where this book is sold or from The Frank R. Walker Company, 5030 N. Harlem Ave., Chicago, IL 60656. Figure 2 shows the top part of this form including the heading. This heading should be filled out on each page of the estimate.

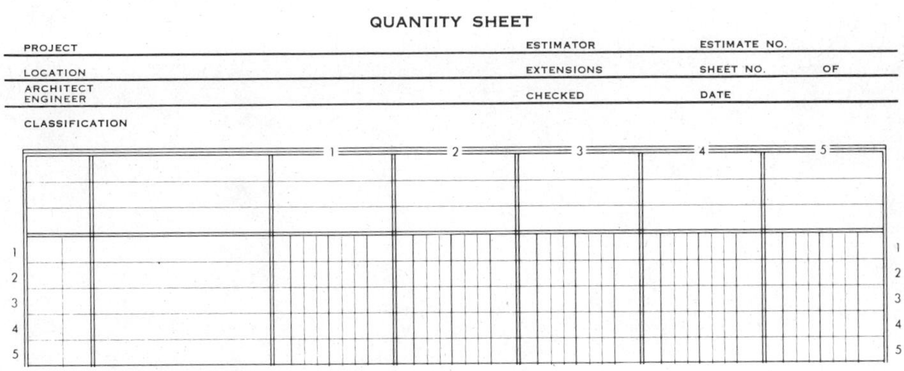
QUANTITY SHEET

PROJECT ESTIMATOR ESTIMATE NO.

LOCATION EXTENSIONS SHEET NO. OF

ARCHITECT ENGINEER CHECKED DATE

CLASSIFICATION

1 2 3 4 5

**FIGURE 2**

The remaining part of this section will follow the UCI logical order except it will begin with Division 3 - Concrete. The reason Division 1 - General Requirements and Division 2 - Site Work are delayed until later is that information from the concrete section can be helpful in determining their quantities.

CHAPTER 5

# DIVISION 3 - CONCRETE

When estimating the quantities and cost of a concrete structure, there are four distinct operations to consider.

1. The quantities and cost of forms, including lumber, nails, tie wire, spacers, together with labor cost erecting and removing forms.

2. The cost of reinforcing steel. In commercial construction the cost is usually established by steel fabrication plants. In residential construction the estimator must establish the quantities and cost of the reinforcing steel, the wire and labor cost for handling and placing.

3. The quantities and cost of concrete mix, together with the cost of labor placing the concrete.

4. The quantities and cost of finishing, curing and protection from the elements.

Some concrete structures will not require all of these operations. For example, a footing for a residence where the soil is stiff enough to be trenched might not need forms. In most cases, concrete structures will require these operations.

The Quantity Sheet is a very flexible form and can be set up to record all types of concrete that occur in a set of drawings. The example in Figure 3 has been set up for a typical residential or small commercial job. The column headed "Description" is used to identify and show the dimensions of the part of the building under consideration. When filling in this column, all dimensions are to be written neatly with a number or a zero before or after the decimal point to make certain that the number is recorded correctly (Example: 0.1 or 1.0 *Not* .1 or 1). In all of the other columns on this page, decimals are rounded off so that 0.01 to 0.49 = 0.0 and 0.5 to 1.0 = 1.0. This policy will save much math and many errors later. The column headed LF (Lineal Foot) is used to estimate the quantities of rebar. The SF (Square Foot) heading is used to estimate the area under a slab requiring fine grading or the area

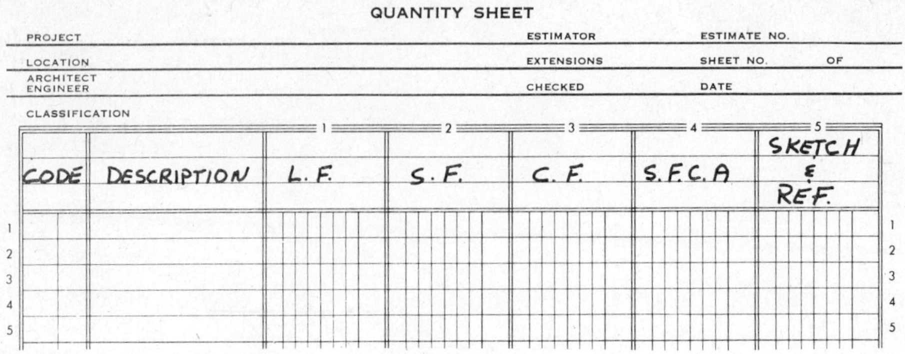

QUANTITY SHEET

PROJECT ESTIMATOR ESTIMATE NO.

LOCATION EXTENSIONS SHEET NO. OF

ARCHITECT ENGINEER CHECKED DATE

CLASSIFICATION

| CODE | DESCRIPTION | L.F. | S.F. | C.F. | S.F.C.A | SKETCH & REF. |
|---|---|---|---|---|---|---|
| | | | | | | |

**FIGURE 3**

requiring curing and rubbing. The CF (Cubic Foot) heading is used to estimate the volume of concrete. SFCA (Square Foot Contact Area) heading is used to estimate the amount of forms required. The description column is used to describe the item that is being taken off, along with the dimensions. The sketch column is used to further describe the item or explain your method of take off. The REF (Reference) is probably the most overlooked column in quantity surveying. This column is very important and should contain the page number and cross section or plan number from which the information was found. The code column is for the division and subdivision numbers.

Now that you have a place for the quantities of each operation, the cost associated with the items must be established. The "Price Block" was developed to provide a place where material quantities can be converted to materials cost and labor cost. The description column within the price block is used to describe the operations of related items. For example, if concrete footings are being estimated, the operations involved in a footing pour are shown in the description column of the price block, Figure 4. The units column is used to state the quantities of each operation. The proper unit for each operation is shown in Appendix A, UCI Order. The columns labeled L and M are used to state the unit cost of labor and material. The unit for concrete is cubic yards. Therefore, a unit cost for concrete is stated in dollars per cubic yards. To complete the final two columns of the price block, you must multiply the number in the unit column by the labor unit cost to give the total labor cost. The figure in the units column is multiplied by the material unit cost to give total material cost.

It cannot be emphasized enough at this point that the unit cost for labor and material must be stated in the same units as those listed in the "units" column. For example, the labor unit cost for placing concrete must be in dollars per cubic yard since the amount of concrete is measured and recorded

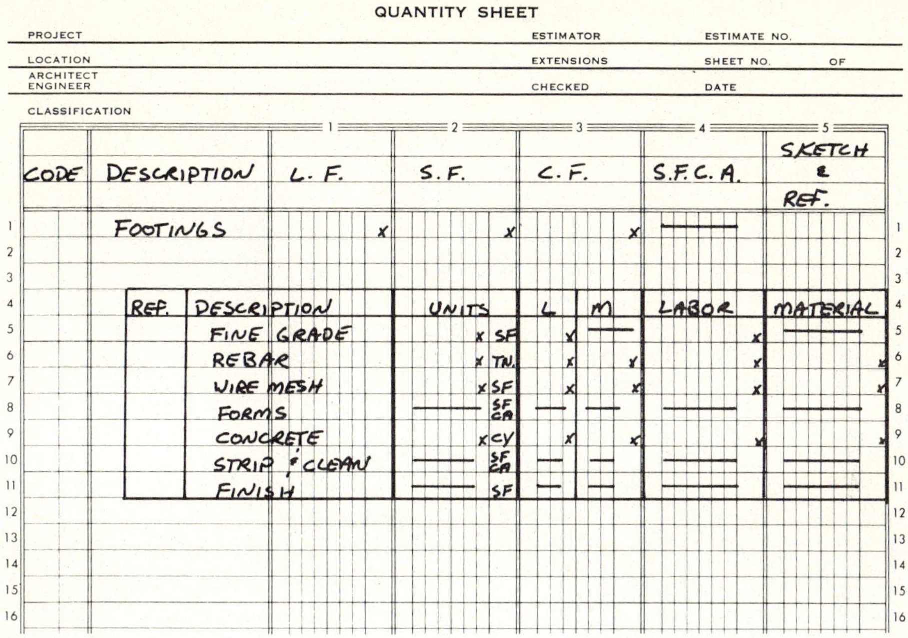

QUANTITY SHEET

PROJECT | ESTIMATOR | ESTIMATE NO.

LOCATION | EXTENSIONS | SHEET NO. OF

ARCHITECT ENGINEER | CHECKED | DATE

CLASSIFICATION

| CODE | DESCRIPTION | L. F. | S. F. | C. F. | S.F.C.A. | SKETCH & REF. |
|---|---|---|---|---|---|---|
| | FOOTINGS | x | x | x | ——— | |

| REF. | DESCRIPTION | UNITS | | L | M | LABOR | MATERIAL |
|---|---|---|---|---|---|---|---|
| | FINE GRADE | x | SF | x | ——— | x | ——— |
| | REBAR | x | TN. | x | x | x | x |
| | WIRE MESH | x | SF | x | x | x | x |
| | FORMS | ——— | SF CA | ——— | ——— | ——— | ——— |
| | CONCRETE | x | CY | x | x | x | x |
| | STRIP & CLEAN | ——— | SF CA | ——— | ——— | ——— | ——— |
| | FINISH | ——— | SF | ——— | ——— | ——— | ——— |

**FIGURE 4**

in cubic yards. Similarly the labor unit cost for finishing concrete must be in dollars per square foot, since the area to be finished is measured and recorded by the square foot.

There are three methods of developing unit costs. The first method depends on the man hours required to produce the units. Using this productivity rate, the estimator can incorporate the local labor and material rates and develop a unit cost. The second method uses crew size. A production rate for an established crew is converted into unit costs using local rates. The last method depends on historical information. The unit cost from past jobs are adjusted to meet the new requirements. The method used in this manual is based on the man hours required to produce the units. The student has no information on productivity; therefore, *The Building Estimator's Reference Book* by Frank R. Walker Company is used as a reference.

By using a price block after each activity, the chance for gross error is reduced. The estimator can make a quick check of the labor and material cost in relation to the quantities. This check can be made by checking the cost of the operation in the price block in relation to the quantity. Then he can check the quantity by checking the calculations in the description column. This format enables this check to be made without loss of time looking in different areas for the information.

When making this check, the estimator must be able to visualize the size of quantities. This visualization of the size of quantities will aid the estimator in determining gross errors in his calculations. Experienced estimators will be able to visualize the size of quantities. The inexperienced estimator can use Table 3 as a guide for the size of common construction quantities.

The most common error is the omission of operations. To minimize the chance of omitting operations, a line should be drawn through columns not used and an X placed in columns requiring calculations. The example of the footing estimate (Figure 4) shows lines through the SFCA column illustrating that no forms are required. A line is also drawn through the fine grade, strip and clean, and finish unit material columns because these items do not require materials. A small x is placed in all columns which are to be used. This x is then erased when the estimator completes the calculation and enters the quantity. This system of lines and x's allows a group of estimators to work on one project and identify what work must be completed. It also makes it easier for one estimator to complete an estimate started by someone else or for that matter to pick up where he left off the day before.

The best way to develop skills in estimating is to work example problems. The following two examples will show methods of quantity take-off and describe how to develop unit prices. The first example is the foundation of a small commercial building followed by a monolithic slab for a single-family residence. To start the estimate for this commercial building foundation, the estimator should study the plans on the following two pages. Then he should set up the proper column titles. (Figure 5.)

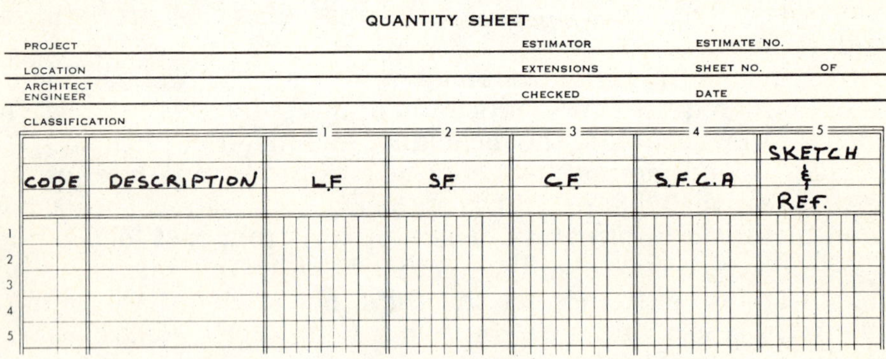
QUANTITY SHEET

PROJECT ESTIMATOR ESTIMATE NO.
LOCATION EXTENSIONS SHEET NO. OF
ARCHITECT ENGINEER CHECKED DATE
CLASSIFICATION

| CODE | DESCRIPTION | 1 L.F. | 2 S.F. | 3 C.F. | 4 S.F.C.A. | 5 SKETCH & REF. |
|---|---|---|---|---|---|---|
| 1 | | | | | | |
| 2 | | | | | | |
| 3 | | | | | | |
| 4 | | | | | | |
| 5 | | | | | | |

**FIGURE 5**

The first activity in the construction of the foundation is the footing. When determining the lineal feet of footing, the dimensions on the foundation plan are building lines; therefore, the dimensions must be adjusted to the center line of the footings. This adjustment should be done on a backup sheet. The

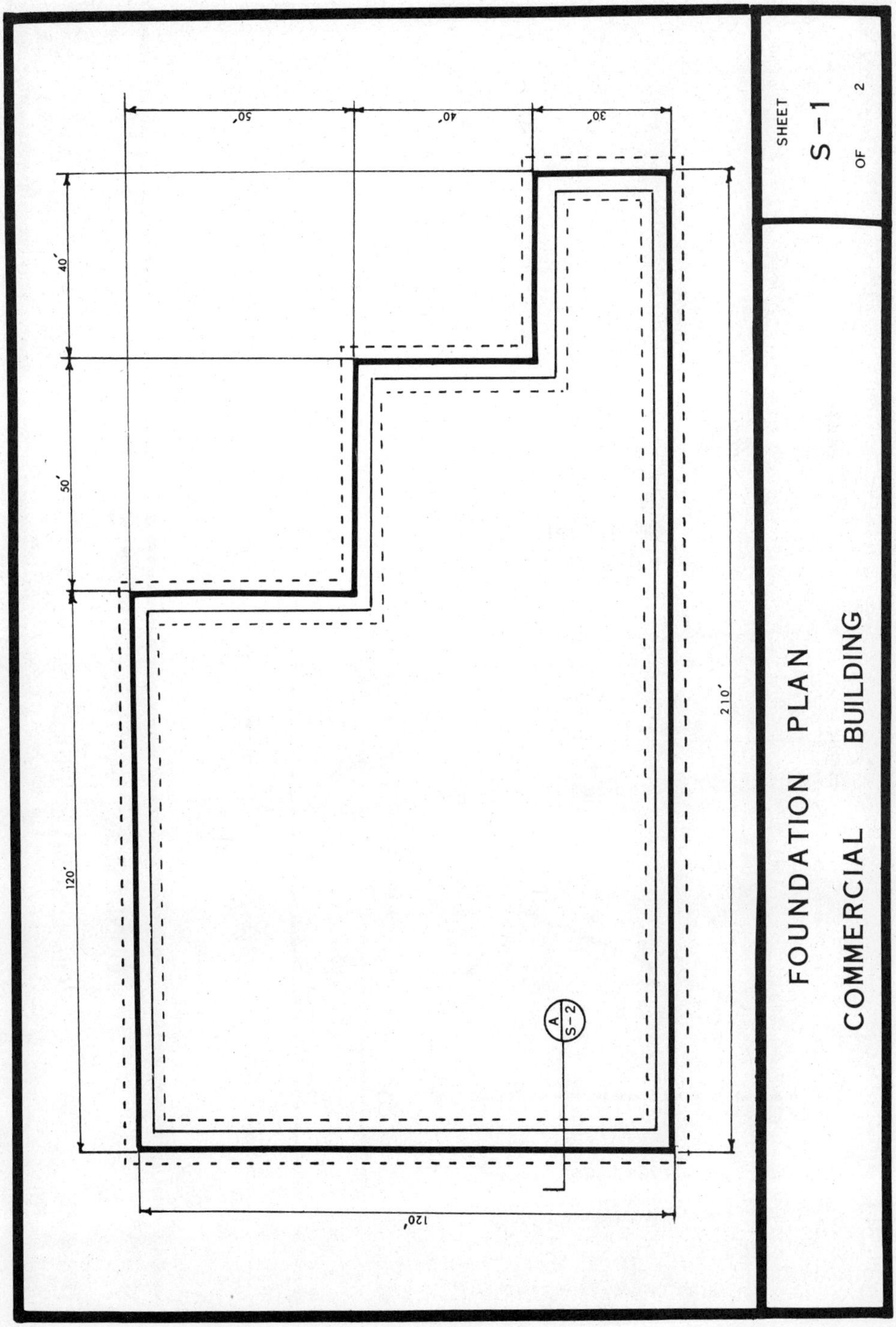
50'
40'
30'
40'
50'
120'
210'
120'
A
S-2
SHEET
S-1
OF 2
FOUNDATION PLAN
COMMERCIAL BUILDING

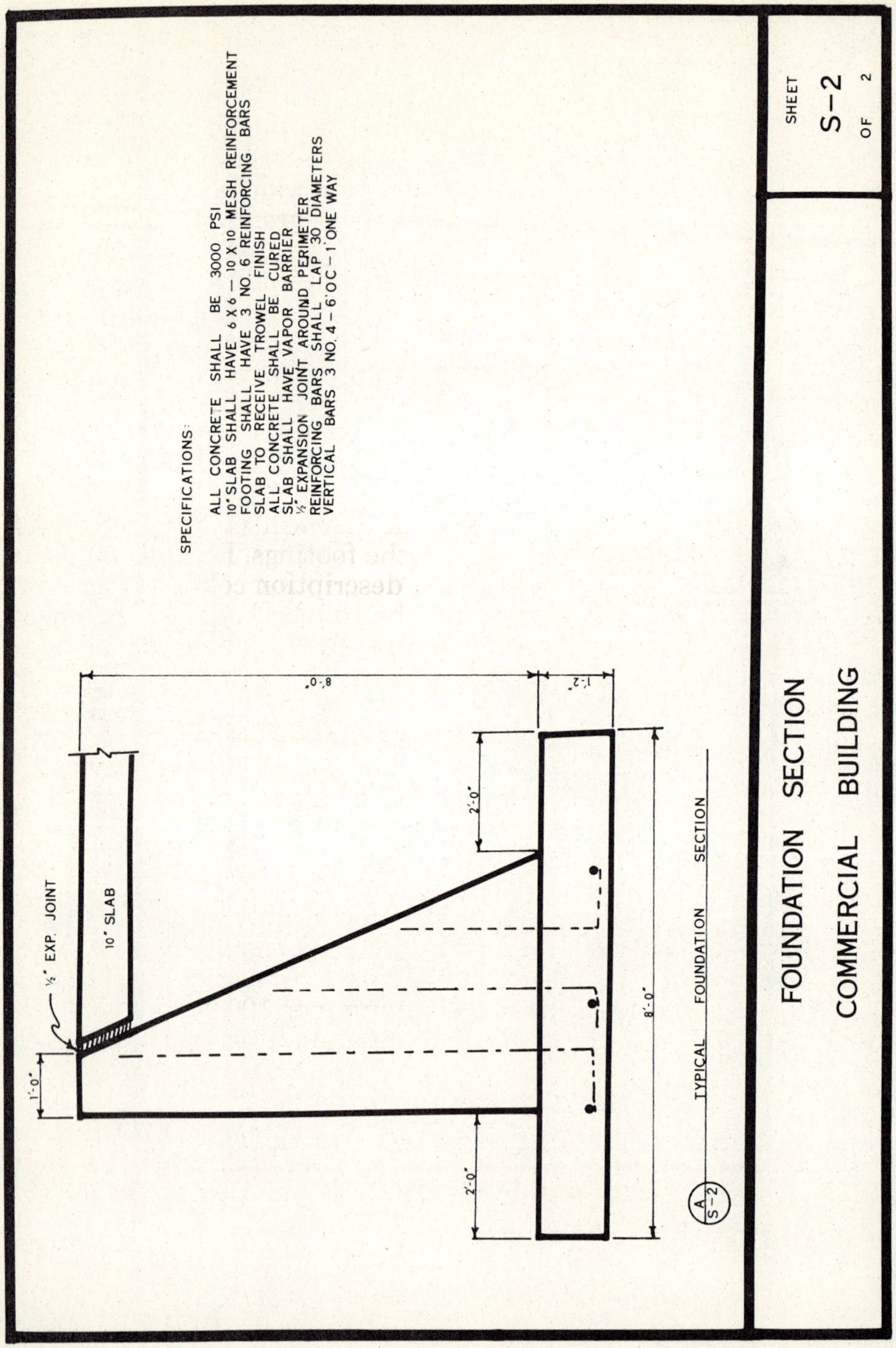
SPECIFICATIONS:
ALL CONCRETE SHALL BE 3000 PSI
10" SLAB SHALL HAVE 6 X 6 — 10 X 10 MESH REINFORCEMENT
FOOTING SHALL HAVE 3 NO. 6 REINFORCING BARS
SLAB TO RECEIVE TROWEL FINISH
ALL CONCRETE SHALL BE CURED
SLAB SHALL HAVE VAPOR BARRIER
½" EXPANSION JOINT AROUND PERIMETER
REINFORCING BARS SHALL LAP 30 DIAMETERS
VERTICAL BARS 3 NO. 4 - 6'OC - 1'ONE WAY
½" EXP. JOINT
10" SLAB
1'-0"
8'-0"
1'-2"
2'-0"
2'-0"
8'-0"
A
S-2
TYPICAL FOUNDATION SECTION
FOUNDATION SECTION
COMMERCIAL BUILDING
SHEET
S-2
OF 2

backup sheet should be set up with a page number and item number to be inserted in the reference column of the estimate. The following portion of a backup sheet shows how to compute the lineal foot of footing. (Figure 6.)

The footing's center line dimension is used to determine the amount of rebar, plus lap. The dimension is also used to determine the square feet of surface area. This area is computed by length x width (644.0' x 8.0' = 5152 $Ft.^2$) and is used to determine the amount of fine grading required. Next, the volume of concrete required is computed by length x width x height (644.0' x 8.0' x 1.17' = 6028 $Ft.^3$). The final quantity to determine is the square feet of contact area for the forms. The footing requires forms on each side. Therefore, the square foot of contact area is computed as follows: length x height x 2 (644.0' x 1.17' x 2 sides = 1507 $Ft.^2$). The information required for the survey is now finished.

Notice in figure 7 that decimal equivelants of inches are used in the description column, but no fractions are used in the L.F., S.F., C.F., or S.F.C.A. Columns.

The next step is to develop a price block that will show the cost of all the operations involved in the construction of the footings. Each of the operations should be listed in proper sequence in the description column. The first operation in a footing pour is to fine grade the soil by hand to provide a level surface. The area that must be graded is the area under the footing. This amount has already been computed in the square foot column in Figure 7. The estimator must now determine a unit labor cost for fine grading. Labor unit cost is usually computed from historical data from past jobs. If the estimator has no historical data, he must depend on reference books to provide cost information. It should take a laborer 6.5 hours to fine grade 100 Sq. Ft. of footing bottom.[6] This production rate is only for fine grading footings. Footings which are dug by machine require large amounts of clean up, squaring, and grade work. In contrast, slab bottoms are usually rolled and compacted in lifts and do not require as much fine grading, thus a higher production rate. The production rates for different operations requiring fine grading are given in the earthwork section of Walker's *Building Estimator's Reference Book.*

Using Walker's production rate of 6.5 hours per 100 Sq. Ft. and the local labor pay scale, the unit cost can be computed on a backup sheet as follows: Figure 8.

The total area requiring fine grading and the labor unit cost can now be inserted into the price block and multiplied to give total labor cost. Fine grading does not require any materials. Therefore, a straight line should be drawn through the material unit cost and total cost columns. Figure 9.

---

[6]Ibid.

| CONCRETE | BACK-UP SHEET | COMMERCIAL BLDG. | 1 |
|---|---|---|---|

| NO | ITEM | TOTAL | REF |
|---|---|---|---|
| 1 | FOOTING - LINEAL FOOT CALCULATION | | S-1 A/S-2 |

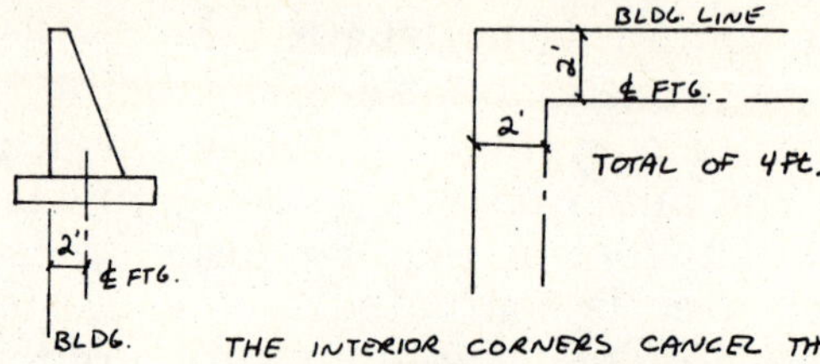


THE INTERIOR CORNERS CANCEL THE EXTERIOR CORNERS 6 EXTERIOR CORNERS − 2 INTERIOR CORNERS LEAVES 4 OFF-SETTING CORNERS.

4 CORNERS X 4 Ft. PER. CORNER = 16 Ft.
660 Ft. (LIN. Ft. BLDG. LINE) − 16 Ft = 644.0Ft.

**FIGURE 6**

QUANTITY SHEET

PROJECT ESTIMATOR ESTIMATE NO.
LOCATION EXTENSIONS SHEET NO. OF
ARCHITECT ENGINEER CHECKED DATE
CLASSIFICATION

| | CODE | DESCRIPTION | L. F. | S. F. | C. F. | S.F.C.A | SKETCH & REF. | |
|---|---|---|---|---|---|---|---|---|
| 1 | | FOOTINGS | | | | | S-1, A/S-2 | 1 |
| 2 | | 644 X 8 X1.17 | 644 | 5152 | 6028 | 1507 | | 2 |
| 3 | | | | | | | | 3 |
| 4 | | | | | | | FORM FTG. | 4 |
| 5 | | | | | | | L.F. B/U 1-1 | 5 |
| 6 | | | | | | | | 6 |
| 7 | | | | | | | | 7 |
| 8 | | | | | | | | 8 |
| 9 | | | | | | | | 9 |
| 10 | | | | | | | | 10 |

**FIGURE 7**

| | | | |
|---|---|---|---|
| 2 | LABOR UNIT COST FOR FINE GRADE | | WALKER P.164 |

| TRADE | HOURS | RATE | TOTAL |
|---|---|---|---|
| LABOR | 6.5 | 4.50 | 29.25 |
| @100 Ft² | | | 29.25 |
| @ Ft² | | | 0.29 |

$0.29

FIGURE 8

QUANTITY SHEET

PROJECT ESTIMATOR ESTIMATE NO.
LOCATION EXTENSIONS SHEET NO. OF
ARCHITECT ENGINEER CHECKED DATE
CLASSIFICATION

| | CODE | DESCRIPTION | L.F. (1) | S.F. (2) | C.F. (3) | S.F.C.A (4) | SKETCH & REF. (5) | |
|---|---|---|---|---|---|---|---|---|
| 1 | | FOOTINGS | | | | | S-1, A/S-2 | 1 |
| 2 | | 644.0×8.0×1.17 | 644 | 5152 | 6028 | 1507 | | 2 |
| 3 | | | | | | | | 3 |
| 4 | | | | | | | FORM FTG. | 4 |
| 5 | | | | | | | L.F. B/U 1-1 | 5 |
| 6 | | | | | | | | 6 |

| | CODE | REF. | DESCRIPTION | UNITS | L | M | LABOR | MATERIAL | |
|---|---|---|---|---|---|---|---|---|---|
| 8 | 202.12 | B/U 1-2 | FINE GRADE | 5152 S.F | 0.29 | — | $ 1494 | — | 8 |
| 9 | | | | | | | | | 9 |
| 10 | | | | | | | | | 10 |
| 11 | | | | | | | | | 11 |
| 12 | | | | | | | | | 12 |
| 13 | | | | | | | | | 13 |
| 14 | | | | | | | | | 14 |
| 15 | | | | | | | | | 15 |

FIGURE 9

The second operation in a footing pour is formwork. From the S.F.C.A. Column, 1507 Sq. Ft. of forms are required for the footings. The estimators must determine a material unit cost and labor unit cost for the forms. The following table from *The Building Estimator's Reference Book* will aid the estimator in the determination.[7]

**Cost of 100 Sq. Ft. of Wall Footing Forms Based on Footings 1'-0" Deep Completely Formed**

| Material | Ft. BM | Rate | Total | Rate | Total |
|---|---|---|---|---|---|
| Sides, 2"x12" planks | 200 | \$.... | \$.... | \$ 0.06 * | \$ 12.00* |
| Stakes and braces | 75 | .... | .... | 0.055* | 4.13* |
| Nails, strap iron, etc. | | .... | .... | 1.20 | 1.20 |
| Placing Forms—25-30 sq. ft. per hr. | **Hours** | | | | |
| Carpenter | 3.50 | .... | .... | 14.60 | 51.10 |
| Labor | 1.75 | .... | .... | 11.85 | 20.74 |
| Stripping Forms—70-75 ft. per hr. | | | | | |
| Labor | 1.50 | .... | .... | 11.85 | 17.78 |
| Cost per 100 sq. ft. | 6.75 | | \$.... | | \$106.95 |
| Material cost per sq. ft. | | | .... | | .17 |
| Labor cost per sq. ft. | | | .... | | .90 |

**TABLE 2**

The preceding Table shows the quantities of materials and production hours required for a one-foot deep footing form in Chicago. The estimator must adjust the material quantities to facilitate a fourteen-inch footing and adjust the material and labor rates for his location. This should be done on a back-up sheet (Figure 10) - then transferred to the price block. Figure 11.

With the cost of the forms established, the estimator should start the next operation, reinforcing steel. The lineal foot of footing from Figure 9 must be adjusted by the required lap of thirty diameters. This new length must then be multiplied by the number of bars in the footing, then converted to tons. The following Table gives the sizes and weights of concrete reinforcing bars.[8] Table 3.

---

[7]Ibid

[8]Ibid

| CONCRETE | BACK-UP SHEET | COMMERCIAL BLDG. | 2 |
|---|---|---|---|

| NO | ITEM | TOTAL | REF. |
|---|---|---|---|
| 1 | UNIT LABOR AND MATERIAL COST FOR FOOTING FORMS | | WALKER P. 217 |

| MATERIAL @ 100 Ft² | Ft. BM | RATE | TOTAL | |
|---|---|---|---|---|
| SIDES 2"X8" PLANKS | 133 | .05* | $6.65 | |
| 2"X6" PLANKS | 100 | .047* | 4.70 | |
| STAKES AND BRACES | 75 | .243 | 18.22 | |
| NAILS | | | 1.20 | |
| TOTAL @ 100 Ft² | | | 30.78 | |
| TOTAL @ Ft² FOR MATERIALS | | | $0.31 | $0.31 |

* BASED ON SIX USES OF LUMBER
2"X8" COST = $.327 PER Ft.BM./6 USES = .05
2"X6" COST = .285 PER FT.BM/6 USES = .047

| LABOR @ 100 Ft² | HOURS | RATE | TOTAL | |
|---|---|---|---|---|
| CARPENTER | 3.50 | 6.67 | $23.35 | |
| LABOR | 1.75 | 4.50 | 7.88 | |
| TOTAL @ 100 Ft² | | | 31.23 | |
| TOTAL @ Ft² FOR LABOR | | | 0.31 | $0.31 |

**FIGURE 10**

DUE: 6-14-80 COMMERCIAL BUILDING PAGE 1 of 1

DESIGNER

OWNER 300 CONCRETE PAT BOURGEOIS

| | CODE | DESCRIPTION | L.F. | S.F. | C.F. | S.F.C.A. | SKETCH & REF. | |
|---|---|---|---|---|---|---|---|---|
| 1 | | FOOTINGS | | | | | S-1 A/S-2 | 1 |
| 2 | | 644.0X 8.0X 1.17 | 644 | 5152 | 6028 | 1507 | | 2 |
| 3 | | | | | | | 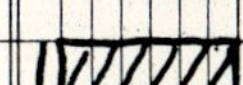 | 3 |
| 4 | | | | | | | FORM FTG. | 4 |
| 5 | | | | | | | LF. B/U 1-1 | 5 |
| 6 | | | | | | | | 6 |

| | CODE | REF | DESCRIPTION | UNITS | L | M | LABOR | MATERIAL | |
|---|---|---|---|---|---|---|---|---|---|
| 8 | 202.12 | B/U 1-2 | FINE GRADE | 5152 SF | 0.29 | — | 1494 | — | 8 |
| 9 | 302.12 | B/U 2-1 | FORM WORK | 1507 SFCA | 0.31 | 0.31 | 467 | 467 | 9 |
| 10 | | | | | | | | | 10 |
| 11 | | | | | | | | | 11 |
| 12 | | | | | | | | | 12 |

**FIGURE 11**

**Standard Sizes and Weights of Concrete Reinforcing Bars**

| Bar Designation (Numbers) | Unit Weight Pounds Per Foot | Nominal Dimensions,—Rounded Sections | | |
|---|---|---|---|---|
| | | Inches Diameter | Cross Sectional area Square Inches | Perimeter Inches |
| 2 | 0.167 | .250 | 0.05 | 0.786 |
| 3 | 0.376 | .375 | 0.11 | 1.178 |
| 4 | 0.668 | .500 | 0.20 | 1.571 |
| 5 | 1.043 | .625 | 0.31 | 1.963 |
| 6 | 1.502 | .750 | 0.44 | 2.356 |
| 7 | 2.044 | .875 | 0.60 | 2.749 |
| 8 | 2.670 | 1.000 | 0.79 | 3.142 |
| 9 | 3.400 | 1.128 | 1.00 | 3.544 |
| 10 | 4.303 | 1.270 | 1.27 | 3.990 |
| 11 | 5.313 | 1.410 | 1.56 | 4.430 |
| 14S | 7.650 | 1.693 | 2.25 | 5.320 |
| 18S | 13.600 | 2.257 | 4.00 | 7.090 |

**TABLE 3**

Reinforcing bars may be purchased from a warehouse in stock lengths, usually twenty feet. This requires all cutting and bending to be done at the job. Typically in commercial work they are purchased cut-to-length and bent, ready to be placed in the building. For large, complex commercial jobs with bar sizes of number six and larger, it is recommended that the reinforcing steel be purchased, bent, bundled and tagged from a fabrication plant. When received from the plant the bars will be accompanied by a shop drawing showing exact placement with each bar tagged to its proper location. For our small commercial building we will estimate the quantities and cost of rebar using twenty foot lengths. The total tons of number six horizontal reinforcing is calculated on the back-up sheet. Figure 12.

The number four vertical reinforcing bars are spaced one foot apart and six feet on center. Because the bars are bought in twenty foot sections, the estimator should try to minimize waste. The following calculations should minimize waste. Figure 14.

The labor unit cost for setting reinforcing bars for an average reinforcing steel setter is computed as follows.[8-A] Figure 13. The unit cost of $122.72 per ton can only be used when it is not necessary to tie the bars in place.

---

[8-A]Ibid

| | | | |
|---|---|---|---|
| 2 | REINFORCING BARS - HORIZONTAL No. 6<br>TONS OF REBAR<br>LAP = 30 DIAMETERS = 30 X .750" = 22.5"/BAR<br>22.5" = 1.87 Ft.<br>BARS MUST LAP 1.87 ft FOR EACH STANDARD 20 Ft. SECTION<br>THIS GIVES A % LAP OF 1.87 ft / 20 ft = 9%<br>644 LIN. Ft. X 3 BARS X 1.09 = 2106 LIN. Ft.<br>2106 LIN. Ft. REQUIRED / 20 Ft. STD. LENGTH = 105.3 BARS<br>MUST ORDER 106 BARS X 20 Ft = 2120 LIN. Ft.<br>2120 LIN. Ft. X 1.502 #/Ft = 3184 # / 2000 #/T. = | 1.59 T | WALKER P. 321 |

**FIGURE 12**

CONCRETE | BACK-UP SHEET | COMMERCIAL BLDG. 3

| No | ITEM | TOTAL | REF. |
|---|---|---|---|
| 1 | REINFORCING BARS<br>LABOR PLACEMENT UNIT COST | | WALKER P. 324 |

| TRADE | HOURS | RATE | TOTAL |
|---|---|---|---|
| STEEL WORKER | 16 | 7.67 | 122.72 |
| TOTAL COST @ TON | | | 122.72 |

TOTAL: 122.72

**FIGURE 13**

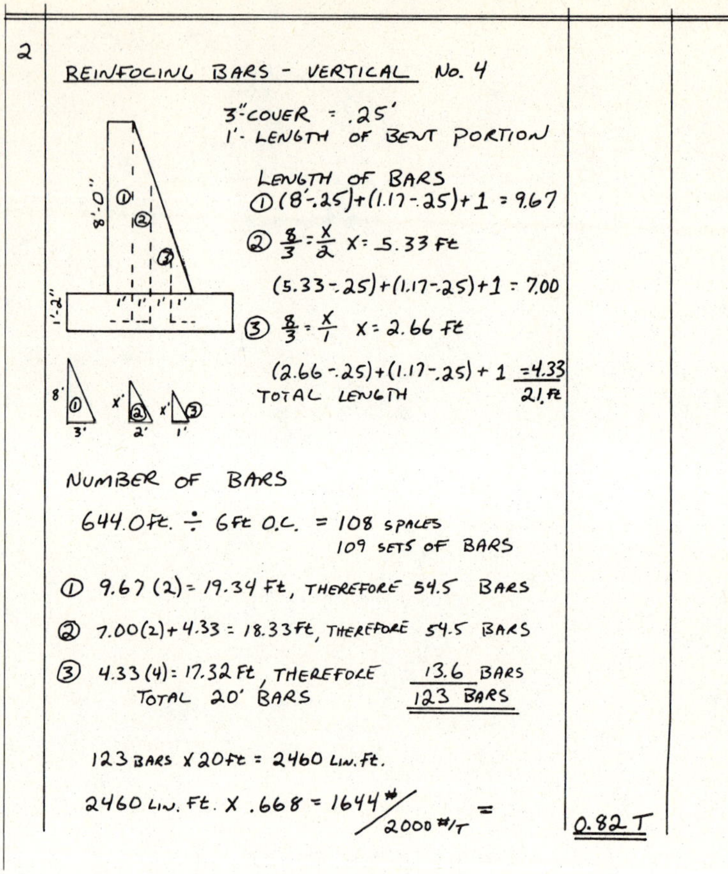
2

REINFOCING BARS - VERTICAL No. 4

3" COVER = .25'
1'- LENGTH OF BENT PORTION

LENGTH OF BARS
① (8'-.25)+(1.17-.25)+1 = 9.67
② 8/3 = X/2 X = 5.33 Ft
(5.33-.25)+(1.17-.25)+1 = 7.00
③ 8/3 = X/1 X = 2.66 Ft
(2.66-.25)+(1.17-.25)+1 = 4.33
TOTAL LENGTH 21. Ft

8'-0" 1'-2" 1' 1' 1' 1' 8' 3' X' 2' X' 1'

NUMBER OF BARS

644.0 Ft. ÷ 6 Ft O.C. = 108 SPACES
109 SETS OF BARS

① 9.67 (2) = 19.34 Ft, THEREFORE 54.5 BARS
② 7.00(2)+4.33 = 18.33 Ft, THEREFORE 54.5 BARS
③ 4.33 (4) = 17.32 Ft, THEREFORE 13.6 BARS
TOTAL 20' BARS 123 BARS

123 BARS X 20 Ft = 2460 LIN. Ft.

2460 LIN. Ft. X .668 = 1644# / 2000#/T = 0.82 T

**FIGURE 14**

The cost per ton for rebar depends on the size of the bar and the number of tons purchased. If the bars are purchased from a fabrication plant, the cost will include all cutting, bending, and usually shipment to the job. If the rebars are purchased in the standard twenty foot lengths, the cost includes only the bars and shipment. In either case, the estimator should call a supplier and get an accurate cost per ton when preparing the estimate. The total tons of reinforcing bars, labor unit cost, and material unit cost can now be included in the price block. Figure 17.

The next operation is the placement of the concrete in the forms. The labor production rate can vary from 0.27 hours per cubic yard to 0.8 hours per cubic yard depending on the means of placement. In the commercial building, the concrete can be placed into the footing directly from the truck. Other operations such as pouring columns, high walls, and multi-story slabs might require a pump or a crane and bucket to place the concrete. The following table shows the various methods of concrete placement and the production rates for concrete foundations.[9] Using the production rate for concrete placed directly into the forms, serviced by two trucks, the labor unit cost is computed as follows: Figure 15.

**Labor Placing One Cu. Yd. of Ready-Mixed Concrete for Concrete Foundations**

| | Hours per Cubic Yard | | |
|---|---|---|---|
| | **Common Labor** | **Hoisting Engineer** | **Fore-man** |
| Method and Quantity Placed per 8-Hr. DayPlacing concrete directly in forms, 1 truck servicing job, 50 to 70 cu. yds. | 0.35-0.50 | .... | 0.13 |
| Placing concrete directly in forms, 2 trucks servicing job, 100 to 150 cu. yds. | 0.27-0.40 | .... | 0.08 |
| Using wheelbarrows, 50 to 70 cu. yds. | 1.25-1.50 | 0.17* | 0.17** |
| Using wheelbarrows, 100 cu. yds and over | 1.00-1.25 | 0.13* | 0.13** |
| Using concrete buggies, 50 to 70 cu. yds. | 1.00-1.25 | 0.17* | 0.17** |
| Using concrete buggies, 100 cu. yds and over | 0.90-1.10 | 0.13* | 0.13** |
| Using 13 cu. ft. power concrete buggies | 0.65-0.75 | 0.13* | 0.13** |

*Where the concrete is hoisted and it is necessary to use a hoisting engineer, add as given in table.

**If union regulations require a mason or cement finisher foreman in charge of crew, add as given in table.

**TABLE 4**

This unit cost is only good where the concrete can be placed directly into the foundation forms, and the labor and foreman rates are the same. The estimator should check with the local ready-mix plants to establish the material unit cost.

The last operation in the placement of the footing is to strip and clean the forms. It takes 1.50 labor hours to strip 100 Sq. Ft. of forms.[10] The labor unit cost to strip the forms is computed as follows: Figure 16.

The complete quantity survey and cost estimate for the footing is shown in Figure 17. This does not include the back up sheets.

---

[9]Ibid

[10]Ibid

| CONCRETE | BACK-UP SHEET | COMMERCIAL BLDG. | 4 |
|---|---|---|---|

| NO. | ITEM | TOTAL | REF. |
|---|---|---|---|
| 1 | LABOR UNIT COST FOR CONCRETE FOUNDATIONS POURED DIRECTLY | $2.01 | WALKER P. 363 |

| TRADE | HOURS | RATE | TOTAL |
|---|---|---|---|
| FORMAN | 0.08 | 6.67 | $0.53 |
| LABOR | 0.33 | 4.50 | 1.48 |
| @ C.Y. | | | $2.01 |

**FIGURE 15**

| NO. | ITEM | TOTAL | REF. |
|---|---|---|---|
| 2 | LABOR UNIT COST FOR STRIPING FORMS | $0.07 | WALKER P. 217 |

| TRADE | HOURS | RATE | TOTAL |
|---|---|---|---|
| LABOR | 1.5 | $4.50 | $6.75 |
| @ 100 Ft. SQ. | | | 6.76 |
| @ Ft. SQ | | | $0.07 |

**FIGURE 16**

QUANTITY SHEET

PROJECT ESTIMATOR ESTIMATE NO.

LOCATION EXTENSIONS SHEET NO. OF

ARCHITECT ENGINEER CHECKED DATE

CLASSIFICATION

| CODE | DESCRIPTION | L.F. | S.F. | C.F. | SFCA | SKETCH & REF |
|---|---|---|---|---|---|---|
| | FOOTINGS | | | | | S-1 |
| | 644.0×8.0×1.17 | 644 | 5152 | 6028 | 1507 |  |
| | | | | | | FORM FTG |

| CODE | REF | DESCRIPTION | UNITS | L | M | LABOR | MATERIAL |
|---|---|---|---|---|---|---|---|
| 202.12 | B/U 1-2 | FINE GRADE | 5152 SF | 0.29 | — | 1494 | — |
| 302.12 | B/U 2-1 | FORM WORK | 1507 SF | 0.31 | 0.31 | 467 | 467 |
| 303.1 | 2-2 3-1 | REBAR | 2.41 TN | 122.7 | 432.00 | 296 | 1041 |
| 301.12 | B/U 4-1 | CONCRETE | 224 CY | 2.01 | 38.50 | 450 | 8624 |
| 302.37 | B/U 4-2 | STRIP & CLEAN | 1507 SF | 0.07 | — | 106 | — |

**FIGURE 17**

The next activity in the construction of the commercial building foundation is the foundation wall. This activity was separated from the footing activity because of the different unit labor and material cost. This difference in unit cost is from the complexity of the forms and method of concrete placement. When setting up the estimate, the estimator should lump only the activities requiring the same unit cost together. The format for the foundation wall is the same as that for the footings. The determination of lineal feet, square feet, cubic feet and square foot contact area should start two to three lines below the price block of the footing activity, if there is room on the same page, otherwise a second sheet should be started.

Because of the irregular shape of the foundation wall, it will be broken down into two simple geometric shapes. The first shape is a rectangle, eight feet by one foot. The linear foot of the first geometric shape must be adjusted in the same manner as the footings. Figure 18.

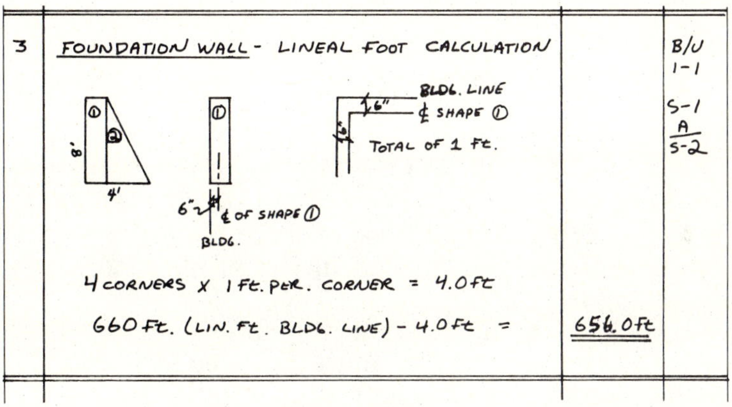


**FIGURE 18**

From the adjusted linear foot of shape one, the volume is computed by Length x Width x Height (656.0 Lin. Ft. x 1.0 Ft. x 8.0 Ft. = 5248 Cubic Ft.) The square foot contact area for forms only includes the area of one side. The remaining square foot contact area will be computed with a second geometric shape. The forms are located on the outside of the wall. Therefore, the lineal foot of building line should be used. The calculation of square foot contact area is computed by Length x Height (660 Lin. Ft. Bldg. Line x 8 Ft. = 5280 Sq. Ft.). The square foot column should be marked through with a straight line because the wall does not require fine grading and the forms will be used to cure the concrete.

The second geometric shape is a triangle eight feet at one leg and three feet at the other. The building line must be adjusted to the centroid of the triangle so the cubic feet and square foot contact area can be computed. Figure 19 shows the adjustment of the building line to the centroid of the triangle.

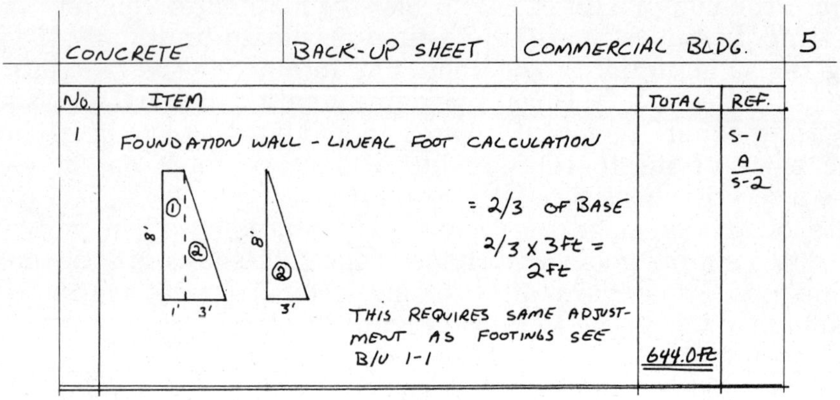


**FIGURE 19**

The 644.0 Ft. is used to compute the volume of concrete, Length x (½ base x height) 644.0 Ft. x (½ x 3.0 x 8.0) = 7728 Cubic Ft. The square foot contact area for forms is Length x Hypotenuse of Triangle 644.0 Ft. x 8.5 = 5502 Sq. Ft. This information should be placed in the proper column below the footing's price block. Figure 20.

The first operation in the foundation wall activity is formwork. The estimator must determine the quantity of material needed to form the foundation wall. The best method of forming the wall is to build two by eight foot panels on the ground, and then erect each panel into position. Figure 21 shows how to calculate the unit material cost for the forms. When calculating the quantity of material required, the estimator must take into account each item required for the forms. These items are then converted to a unit material cost per square foot. This calculation involves the aggregation and adjustment of several tables in *The Building Estimator's Reference Book.* When estimating the unit material cost of forms for slabs, columns and beams, the estimator must build the forms on paper. He should use historical data or reference books to help determine the quantities of plywood, shore wales, column clamps, etc.

The unit labor cost to build and erect the forms must now be calculated. The cost to build the two x eight foot panels must be separated from the cost to erect the forms because the labor cost to build the panels is used six times. The unit labor cost for the building and erecting of the forms is shown in Figure 22.

QUANTITY SHEET

| PROJECT | ESTIMATOR | ESTIMATE NO. | |
|---|---|---|---|
| LOCATION | EXTENSIONS | SHEET NO. | OF |
| ARCHITECT ENGINEER | CHECKED | DATE | |

CLASSIFICATION

| Code | Description | 1 L.F. | 2 S.F. | 3 C.F. | 4 SFCA | 5 Sketch & Ref. |
|---|---|---|---|---|---|---|
| | Footings | | | | | S-1 A/S-2 |
| | 644.0×8.0×1.17 | 644 | 5152 | 6028 | 1507 | |
| | | | | | | Form Ftg. B/U 1-1 |

| Code | Ref | Description | Units | L | M | Labor | Material |
|---|---|---|---|---|---|---|---|
| 202.12 | B/U 1-2 | Fine Grade | 5152 SF | 0.29 | — | 1494 | — |
| 302.12 | B/U 2-1 | Formwork | 1507 SF | 0.31 | 0.31 | 467 | 467 |
| 303.1 | 2-2 3-1 | Rebar | 2.41 Ton | 122.70 | 432.00 | 296 | 1041 |
| 301.12 | B/U 4-1 | Concrete | 224 CY | 2.01 | 38.50 | 450 | 8624 |
| 302.37 | B/U 4-2 | Strip & Clean | 1507 SF | 0.07 | — | 105 | — |

| Code | Description | L.F. | S.F. | C.F. | SFCA | Sketch & Ref. |
|---|---|---|---|---|---|---|
| | Foundation Wall | | | | | |
| | 656.0×1.0×8.0 | 656 | — | 5248 | 5248 | B/U 4-3 |
| | 644.0×3.0×8.0×0.5 | 644 | — | 7728 | 5502 | B/U 5-1 |
| | Totals | 1300 | — | 12976 | 10750 | |

**FIGURE 20**

CONCRETE | BACK UP SHEET | COMMERCIAL BLDG | 6

| No. | ITEM | TOTAL | REF. |
|---|---|---|---|
| 1. | UNIT MATERIAL COST FOR WALL FORM<br>LUMBER COST – 1 2'x8' PANEL | | |

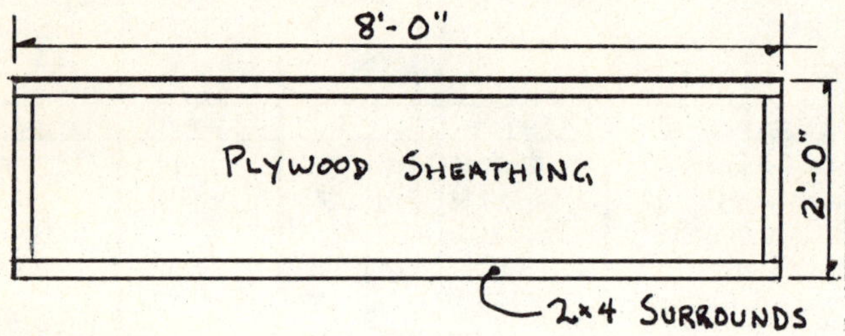


| ITEM | QUANTITY | RATE | TOTAL |
|---|---|---|---|
| 2x4 - 20 LF | 13.3 BF | $0.243 | $3.24 |
| PLYFORM | ½ PC | 19.34 | 9.67 |
| NAILS | — | L/S | 0.50 |
| WALES, 2-2x4-32" O.C. | 8.0 BF | 0.243 | 1.95 |
| COST PER PANEL | | | $15.36 |

REF.: WALKER P. 223

ASSUME 6 USES 15.36/6 — $2.56

$2.56/16 S.F. — $0.25/S.F.
SNAP TIES $0.30 EA - 2'0" O.C — 0.08/S.F.
NAILS & EXPENDABLE BRACING — 0.02/S.F.

REF.: WALKER P. 231

TOTAL UNIT PRICE FOR FORM PANELS — TOTAL: $0.25 PER S.F.

**FIGURE 21**

2 UNIT LABOR COST TO BUILD & ERECT FORMS — WALKER P. 238

LABOR MAKING 100 SQ.FT OF PANEL FORMS

| TRADE | HOURS | RATE | TOTAL |
|---|---|---|---|
| CARPENTER | 3.5 | $6.67 | $23.34 |
| COST PER 100 SQ FT | | | $23.34 |
| COST PER SQFT | | | $0.23 |

LABOR ERECTING 100 SQ.FT OF PANEL FORMS

| TRADE | HOURS | RATE | TOTAL |
|---|---|---|---|
| CARPENTER | 4.5 | $6.67 | $30.02 |
| LABOR | 2.25 | 4.50 | 10.13 |
| COST PER 100 SQ. FT. | | | 40.15 |
| COST PER SQ.FT | | | $0.40 |

BASED ON SIX REUSES OF PANELS
TOTAL UNIT LABOR COST

($0.23 ÷ 6 USES) + $0.40 = $0.44 — $0.44 PER SQ FT

**FIGURE 22**

The second operation in the foundation wall activity is the placement of concrete into the forms. The concrete has to be placed into a one foot opening, nine feet-two inches above the ground. This requires a small crane and bucket to place the concrete. Figure 23 shows the labor unit cost for concrete placement using a crane and bucket.

The labor unit cost takes into account the crane operator but not the rental of the crane. The crane rental cost cannot be included in the labor unit cost because social security and workman's compensation tax will be added to total labor cost later. The crane rental unit cost will be treated as a material unit cost. The last operation in the foundation wall activity is to strip and clean the forms. The unit labor cost will be higher than that for the footings because the wall forms are more complex. It take 2.25 labor hours to strip and clean 100 square feet of the wall panels.[11] Figure 25. Figure 27 shows the quantity survey and cost estimate through the foundation wall activity. This figure does not include the back-up sheets.

[11]Ibid

| CONCRETE | BACK-UP SHEET | COMMERCIAL BLDG. | 7 |
|---|---|---|---|

| No. | ITEM | TOTAL | REF. |
|---|---|---|---|
| 1 | UNIT LABOR COST FOR PLACING CONCRETE INTO FOUNDATION WALL FORMS USEING A CRANE AND BUCKET. | $2.61 | WALKER P. 364 |

| TRADE | HOURS | RATE | TOTAL |
|---|---|---|---|
| LABOR | 0.40 | $4.50 | 1.8 |
| CRANE OP. | 0.05 | 9.60 | .48 |
| FOREMAN | 0.05 | 6.67 | .33 |
| | | | $2.61 |

**FIGURE 23**

| No. | ITEM | TOTAL | REF. |
|---|---|---|---|
| 2 | UNIT CRANE RENTAL COST FOR FOUNDATION<br>QUANTITY PLACED PER DAY 150 TO 170 cu.yd.<br>481 cu.yd IN FND. WALL ÷ 160 cu.yd. PER DAY = 3 DAYS | $0.62 | WALKER P. 364 |

| EQUIPMENT | DAYS | RATE | TOTAL |
|---|---|---|---|
| CRANE | 3 | $100.00 | $300.00 |
| @ 481 cu.yd. | | | $300.00 |
| @ cu.yd. | | | 0.62 |

**FIGURE 24**

| No. | ITEM | TOTAL | REF. |
|---|---|---|---|
| 3 | UNIT LABOR COST FOR STRIPING FORMS | $0.10 | WALKER P. 238 |

| TRADE | HOURS | RATE | TOTAL |
|---|---|---|---|
| LABOR | 2.25 | $4.50 | $10.13 |
| COST @ 100 SQ.Ft | | | 10.13 |
| COST @ SQ.Ft | | | $0.10 |

**FIGURE 25**

The pouring of the slab is the last activity in the foundation of the commercial building. The operations required for this activity are fine grade, vapor barrier, wire mesh, expansion joint, screeds, concrete, finishing and curing. The square foot and cubic foot columns must be completed to provide information for these operations. The lineal foot and square foot contact area columns are not needed and should be marked through with a straight line.

The easiest way to compute the square foot of slab area is to break the slab into three simple rectangles. The total square foot area and the volume of the slab is then the sum of the three sections. The following back-up sheet shows how this is done. Figure 26. The volume and surface area are inserted onto the accounting paper in the proper columns. Figure 26.

| CONCRETE | BACK-UP SHEET | COMMERCIAL BLDG. | 8 |
|---|---|---|---|

| NO. | ITEM | TOTAL | REF. |
|---|---|---|---|
| 1. | AREA OF SLAB | | S-1 A S-2 |

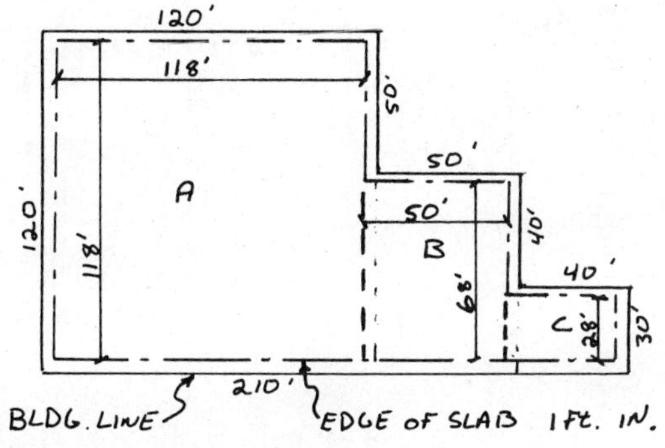


| | | | |
|---|---|---|---|
| A. | 118' X 118' = | 13,924 | |
| B. | 50' X 68' = | 3,400 | |
| C | 28' X 40' = | 1,120 | |
| | TOTAL | 18,447 SQ.Ft. | 18,447 SQ.Ft. |

**FIGURE 26**

QUANTITY SHEET

| PROJECT | ESTIMATOR | ESTIMATE NO. | |
|---|---|---|---|
| LOCATION | EXTENSIONS | SHEET NO. | OF |
| ARCHITECT ENGINEER | CHECKED | DATE | |
| CLASSIFICATION | | | |

| CODE | DESCRIPTION | L.F. | S.F. | C.F. | S.F.C.A. | SKETCH & REF. |
|---|---|---|---|---|---|---|
| | FOOTINGS | | | | | S-1 A/S-2 |
| | 644.0 × 8.0 × 1.17 | 644 | 5152 | 6028 | 1507 | |
| | | | | | | FORM FTG |
| | | | | | | B/U 1-1 |

| CODE | REF. | DESCRIPTION | UNITS | L | M | LABOR | MATERIAL |
|---|---|---|---|---|---|---|---|
| 202.12 | B/U 1-2 | FINE GRADE | 5152 SF | 0.29 | — | 1494 | — |
| 302.12 | B/U 3-1 | FORMWORK | 1507 SF | 0.31 | 0.31 | 467 | 467 |
| 303.1 | 2-2 3-1 | REBAR | 2.41 TNS | 122.70 | 432.00 | 296 | 1041 |
| 301.12 | B/U 4-1 | CONCRETE | 224 CY | 2.01 | 38.50 | 450 | 8624 |
| 302.37 | B/U 4-2 | STRIP & CLEAN | 1507 SF | 0.07 | — | 105 | — |

| CODE | DESCRIPTION | L.F. | S.F. | C.F. | S.F.C.A. | SKETCH & REF. |
|---|---|---|---|---|---|---|
| | FOUNDATION WALL | | | | | |
| | 656.0 × 1.0 × 8.0 | 656 | — | 5248 | 5248 | B/U 4-3 |
| | 644.0 × 3.0 × 8.0 × 0.5 | 644 | — | 7728 | 5502 | B/U 5-1 |
| | TOTALS | 1300 | — | 12976 | 10750 | |

| CODE | REF. | DESCRIPTION | UNITS | L | M | LABOR | MATERIAL |
|---|---|---|---|---|---|---|---|
| 302.13 | S-2 6-1 | FORMWORK | 10750 SF | 0.44 | 0.25 | 4744 | 2688 |
| 301.13 | B/U 7-1 | CONCRETE | 481 CY | 2.61 | 38.50 | 1225 | 18579 |
| | B/U 7-3 | CRANE RENTAL | 3 DAY | — | 100.00 | — | 300 |
| 302.37 | B/U 7-3 | STRIP & CLEAN | 10750 SF | 0.10 | — | 1075 | — |

**FIGURE 27**

The first operation in the slab activity is fine grading. The fill under the slab is applied in lifts and usually rolled or compacted. These operations will be considered with earthwork. The fine grading for the slab is usually considered part of the slab activity. It takes 1.5 labor hours per 100 square feet to fine grade for slabs.[12] Using this production rate, the unit labor cost for fine grading is computed in Figure 28.

Polyethylene vapor barriers usually come in 100' rolls with widths ranging from two feet to twenty feet. The vapor barrier should have a lap of ten percent. To account for the ten percent, the square foot of floor area should be

[12] Ibid

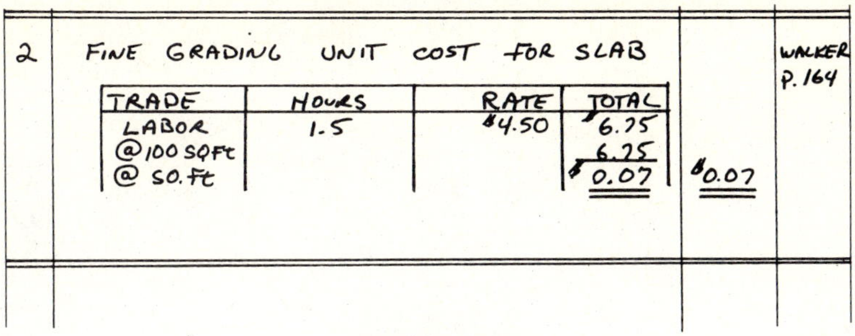


FIGURE 28

increased (18,444 Sq. Ft. x 1.10 = 20,288 Sq. Ft.). The unit material cost for the polyethylene is $0.01 per square foot.[13] The vapor barrier is usually placed along with the welded wire mesh with no unit labor cost associated with its placement. The unit labor cost and the increased square foot area are included in the price block. Figure 30.

The next operation in the slab activity is the welded wire mesh. The mesh must be lapped by ten percent. Therefore, the same increase in square foot area for the vapor barrier should be inserted into the units column of the price block. The unit material cost for 6 x 6 - 10 x 10 welded wire mesh can be received from material suppliers. The wire mesh usually costs about $9.25 per 100 square feet, giving a unit material cost of $0.09 per square foot. It should take a laborer 0.53 hours to lay 100 square feet of mesh. Figure 29 shows the calculation of unit labor cost for laying 100 S.F. reinforcing wire mesh.

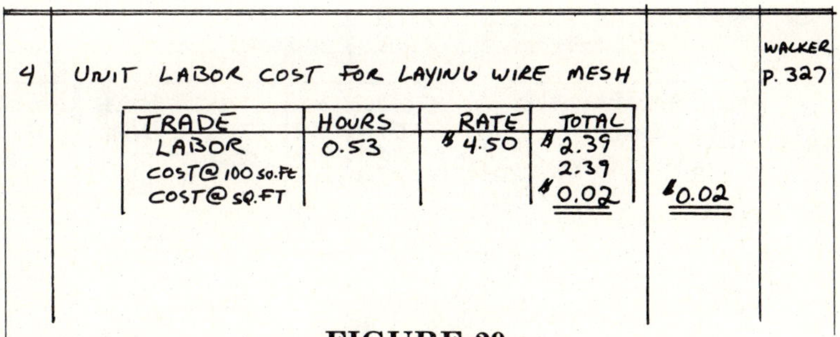


FIGURE 29

[13]Ibid

## QUANTITY SHEET

PROJECT | ESTIMATOR | ESTIMATE NO.

LOCATION | EXTENSIONS | SHEET NO. OF

ARCHITECT ENGINEER | CHECKED | DATE

CLASSIFICATION

| Code | Description | 1 L.F. | 2 S.F. | 3 C.F. | 4 S.F.C.A. | 5 Sketch & Ref. |
|---|---|---|---|---|---|---|
| | **Footings** | | | | | S-1 A/2-2 |
| | 644.0×8.0×1.17 | 644 | 5152 | 6028 | 1507 | |
| | | | | | | Form Ftg |
| | | | | | | B/U 1-1 |

| Code | Ref. | Description | Units | L | M | Labor | Material |
|---|---|---|---|---|---|---|---|
| 202.12 | B/U 1-2 | Fine Grade | 5152 SF | 0.29 | — | 1494 | — |
| 302.12 | B/U 2-1 | Formwork | 1507 SF | 0.31 | 0.31 | 467 | 467 |
| 303.1 | 2-2 3-1 | Rebar | 2.41 Tns | 122.70 | 432.00 | 296 | 1041 |
| 301.12 | B/U 4-1 | Concrete | 224 CY | 2.01 | 38.50 | 450 | 8624 |
| 302.37 | B/U 4-2 | Strip & Clean | 1507 SF | 0.07 | — | 105 | — |

| Code | Description | 1 L.F. | 2 S.F. | 3 C.F. | 4 S.F.C.A. | 5 Sketch & Ref. |
|---|---|---|---|---|---|---|
| | **Foundation Wall** | | | | | |
| | 656.0×1.0×8.0 | 656 | — | 5248 | 5248 | B/U 4-3 |
| | 644.0×3.0×8.0×0.5 | 644 | — | 7728 | 5502 | B/U 5-1 |
| | Totals | 1300 | — | 12976 | 10750 | |

| Code | Ref. | Description | Units | L | M | Labor | Material |
|---|---|---|---|---|---|---|---|
| 302.13 | S-2 6-1 | Formwork | 10750 SF | 0.44 | 0.25 | 4744 | 2688 |
| 301.13 | B/U 7-1 | Concrete | 481 CY | 2.61 | 38.50 | 1225 | 18579 |
| | B/U 7-3 | Crane Rental | 3 Day | — | 100.00 | — | 300 |
| 302.37 | B/U 7-3 | Strip & Clean | 10750 SF | 0.10 | — | 1075 | — |

| Code | Description | 1 L.F. | 2 S.F. | 3 C.F. | 4 S.F.C.A. | 5 Sketch & Ref. |
|---|---|---|---|---|---|---|
| | **Slab** | | | | | |
| | A 118.0×118.0×0.83 | — | 13924 | 11556 | — | B/U 8-1 |
| | B 50.0×68.0×0.83 | — | 3400 | 2822 | — | |
| | C 28.0×40.0×0.83 | — | 1120 | 930 | — | |
| | Totals | — | 18444 | 15308 | — | |

| Code | Ref. | Description | Units | L | M | Labor | Material |
|---|---|---|---|---|---|---|---|
| 202.12 | B/U 8-2 | Fine Grade | 18444 SF | 0.07 | — | 1291 | — |
| | | Vapor Barrier | 20288 SF | — | 0.01 | — | 203 |
| 303.2 | B/U 7-4 | Wire Mesh | 20288 SF | 0.02 | 0.09 | 406 | 1826 |
| 303.3 | | Expansion Joint | 656 LF | 0.01 | 0.20 | 7 | 131 |
| | | Crane Rental | 3 Day | — | 300.00 | — | 300 |
| 301.18 | | Concrete | 567 CY | 2.61 | 38.50 | 1480 | 21829 |
| 303.24 | B/U 9-1 | Finish & Curing | 18444 SF | 0.05 | 0.02 | 922 | 368 |
| | | Page Total → | | | | 13962 | 56356 |

**FIGURE 30**

The placement of the expansion joint is the next operation. The expansion joint is taken off in lineal feet with the labor and material unit cost per lineal foot. The unit material cost is usually provided by the material suppliers. The unit labor cost is usually calculated from historical cost. A unit material cost of $0.20 and a unit labor cost of $0.01 will be used for the commercial building.

The unit labor cost for placing the concrete in the slab will depend on the method of placement. The estimator must establish the best and most economical method that will suit the building's needs. For the commercial building, a crane and bucket will be used for the foundation wall and is probably the best method for pouring a slab of this size. The unit labor and material cost have already been shown in Figure 27.

Finishing and curing are the last operations required in the slab activity. Both of these operations are based on square foot of the area of slab. Figure 31 shows how to compute the labor and material unit cost.

| CONCRETE | BACK-UP SHEET | | COMMERCIAL BLDG. | | | | 9 |
|---|---|---|---|---|---|---|---|
| NO | ITEM | | | | | TOTAL | REF. |
| 1 | | UNIT LABOR AND MATERIAL COST FOR FINISHING AND CURING | | | | | WALKER 398 |
| | LABOR | TRADE | HOURS | RATE | TOTAL | | |
| | | MASON OPERATOR | 8 | $7.58 | $60.64 | | |
| | | MASON | 16 | 7.58 | 121.28 | | |
| | | COST @ 4,000 SQ.FT | | | 181.92 | | |
| | | COST @ SQ.FT. | | | $0.05 | $0.05 | |
| | MATERIAL | ITEM | NO | RATE | TOTAL | | WALKER P. 449 |
| | | MACHINE | 8 | $7.00 | $40.00 | | |
| | | CURING COMPOUND | 12 GAL. | $1.75 | 21.00 | | |
| | | COST @ 4,000 SQ.FT. | | | 61.00 | | |
| | | COST @ SQ.FT. | | | 0.02 | $0.02 | |

**FIGURE 31**

The complete take-off and cost estimate along with all the required back-up sheets are shown on page 41 through 50. The student should note that the code column shows the uniform construction index's cost code. This code appears to the left of the price block and is used to code the different operations to be used later for cost control reports. The reference column was used to show the plan page number or back-up sheet number associated with the quantity survey. The concrete estimate for the small commercial building requires three activities; footings, foundation wall and slab. These activities were not considered as one because they require different unit costs. The reference column within the price block shows the number of the back-up sheet where lap, unit labor, and unit material costs were calculated.

The example of the commercial building shows the proper format for concrete estimating. The example shows how the estimator can build the building on paper in the sequence of construction activities. After each activity a price block is developed showing the cost of each activity. This format allows the estimate to be used as the building is being constructed to determine crew sizes, cash flow and predict cost overruns.

This example does not show all of the many different activities that may be required for a concrete building. The example only covered footings, walls, and slabs on grade. The estimator will be required to estimate concrete walks, columns, beams, suspended slabs, etc. Because of its flexibility the format will work for all concrete activities. Figures 21 and 22 show how to compute the unit material and labor cost for the foundation wall forms. The estimator can use this piece by piece method to develop unit labor and material costs for concrete columns, beams or suspended slabs.

When all of the concrete costs have been calculated for the building, a total labor cost and a total material cost is put at the bottom of the last page of the concrete estimate. These costs are later carried forward to the summary sheet.

The second example of a monolithic slab for a single-family residence shown on pages 51 and 52 will be used as a practice problem. The problem contains only the slab activity requiring one price block. The problem should be set up on quantity sheets with the proper headings. The plans should be studied, dividing the structure into simple geometric shapes. Figure 32. From these simple geometric shapes, the lineal foot of grade beam and thickened slab, the surface area, the volume of concrete and the square foot contact area can be calculated. The operations included in the slab activity are: fine grade formwork, vapor barrier, rebar, wire mesh, concrete, finishing, curing and hand rubbing. The estimator should prepare a complete take-off and cost estimate, including back-up sheets, showing the method of figuring unit costs.

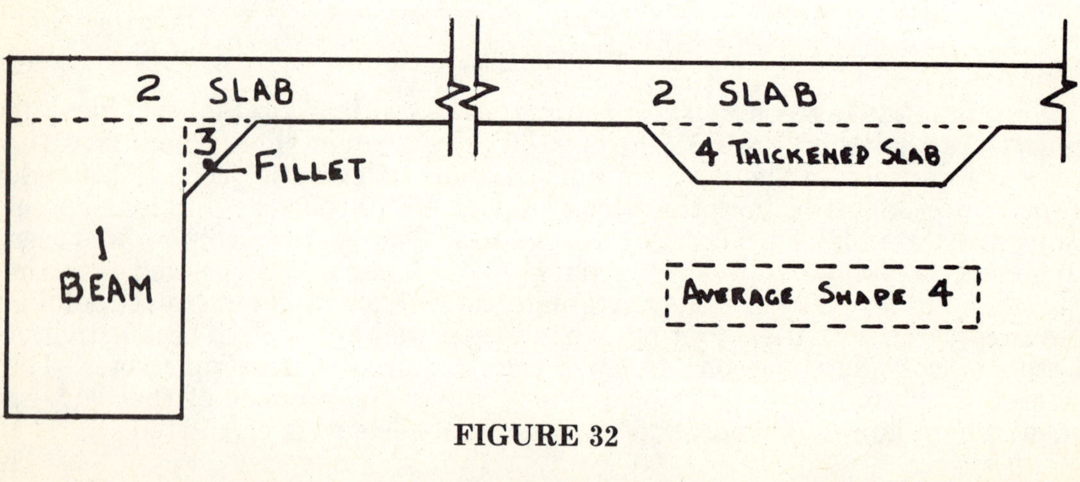


**FIGURE 32**

QUANTITY SHEET

PROJECT | ESTIMATOR | ESTIMATE NO.
LOCATION | EXTENSIONS | SHEET NO. OF
ARCHITECT ENGINEER | CHECKED | DATE

CLASSIFICATION

| CODE | DESCRIPTION | L.F. | S.F. | C.F. | S.F.C.A. | SKETCH & REF. |
|---|---|---|---|---|---|---|
| | FOOTINGS | | | | | S-1 A/S-2 |
| | 644.0 x 8.0 x 1.17 | 644 | 5152 | 6028 | 1507 | |
| | | | | | | FORM FTG |
| | | | | | | B/U 1-1 |

| CODE | REF. | DESCRIPTION | UNITS | L | M | LABOR | MATERIAL |
|---|---|---|---|---|---|---|---|
| 202.12 | B/U 1-2 | FINE GRADE | 5152 SF | 0.29 | — | 1494 | — |
| 302.12 | B/U 2-1 | FORMWORK | 1507 SF | 0.31 | 0.31 | 467 | 467 |
| 303.1 | 2-2 3-1 | REBAR | 2.41 TNS | 122.70 | 432.00 | 296 | 1041 |
| 301.12 | B/U 4-1 | CONCRETE | 224 CY | 2.01 | 38.50 | 450 | 8624 |
| 302.37 | B/U 4-2 | STRIP & CLEAN | 1507 SF | 0.07 | — | 105 | — |

| CODE | DESCRIPTION | L.F. | S.F. | C.F. | S.F.C.A. | SKETCH & REF. |
|---|---|---|---|---|---|---|
| | FOUNDATION WALL | | | | | |
| | 656.0 x 1.0 x 8.0 | 656 | — | 5248 | 5248 | B/U 4-3 |
| | 644.0 x 3.0 x 8.0 x 0.5 | 644 | — | 7728 | 5502 | B/U 5-1 |
| | TOTALS | 1300 | — | 12976 | 10750 | |

| CODE | REF. | DESCRIPTION | UNITS | L | M | LABOR | MATERIAL |
|---|---|---|---|---|---|---|---|
| 302.13 | S-2 6-1 | FORMWORK | 10750 SF | 0.44 | 0.25 | 4744 | 2688 |
| 301.13 | B/U 7-1 | CONCRETE | 481 CY | 2.61 | 38.50 | 1225 | 18579 |
| | B/U 7-3 | CRANE RENTAL | 3 DAY | — | 100.00 | — | 300 |
| 302.37 | B/U 7-3 | STRIP & CLEAN | 10750 SF | 0.10 | — | 1075 | — |

| CODE | DESCRIPTION | L.F. | S.F. | C.F. | S.F.C.A. | SKETCH & REF. |
|---|---|---|---|---|---|---|
| | SLAB | | | | | |
| | A 118.0 x 118.0 x 0.83 | — | 13924 | 11556 | — | B/U 8-1 |
| | B 50.0 x 68.0 x 0.83 | — | 3400 | 2822 | — | |
| | C 28.0 x 40.0 x 0.83 | — | 1120 | 930 | — | |
| | TOTALS | — | 18444 | 15308 | — | |

| CODE | REF. | DESCRIPTION | UNITS | L | M | LABOR | MATERIAL |
|---|---|---|---|---|---|---|---|
| 202.12 | B/U 8-2 | FINE GRADE | 18444 SF | 0.07 | — | 1291 | — |
| | | VAPOR BARRIER | 20288 SF | — | 0.01 | — | 203 |
| 303.2 | B/U 7-4 | WIRE MESH | 20288 SF | 0.02 | 0.09 | 406 | 1826 |
| 303.3 | | EXPANSION JOINT | 656 LF | 0.01 | 0.20 | 7 | 131 |
| | | CRANE RENTAL | 3 DAY | — | 300.00 | — | 300 |
| 301.18 | | CONCRETE | 567 CY | 2.61 | 38.50 | 1480 | 21829 |
| 303.24 | B/U 9-1 | FINISH & CURING | 18444 SF | 0.05 | 0.02 | 922 | 368 |
| | | PAGE TOTAL → | | | | 13962 | 56356 |

| CONCRETE | BACK-UP SHEET | COMMERCIAL BLDG. | 1 |
|---|---|---|---|

| NO | ITEM | TOTAL | REF |
|---|---|---|---|
| 1 | FOOTING - LINEAL FOOT CALCULATION | | S-1 A / S-2 |

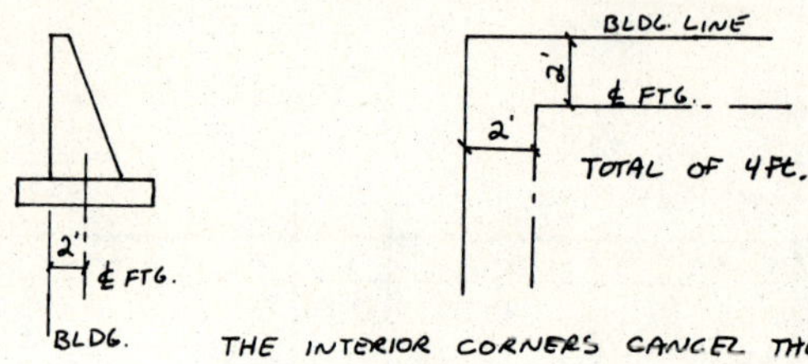


THE INTERIOR CORNERS CANCEL THE EXTERIOR CORNERS 6 EXTERIOR CORNERS - 2 INTERIOR CORNERS LEAVES 4 OFF-SETTING CORNERS.

4 CORNERS X 4 Ft. PER. CORNER = 16 Ft.
660 Ft. (LIN. Ft. BLDG. LINE) - 16 Ft = **644.0 ft.**

| NO | ITEM | TOTAL | REF |
|---|---|---|---|
| 2 | LABOR UNIT COST FOR FINE GRADE | $0.29 | WALKER P. 164 |

| TRADE | HOURS | RATE | TOTAL |
|---|---|---|---|
| LABOR | 6.5 | 4.50 | 29.25 |
| @ 100 ft² | | | 29.25 |
| @ Ft² | | | 0.29 |

| CONCRETE | BACK-UP SHEET | COMMERCIAL BLDG. | 2 |
|---|---|---|---|

| NO | ITEM | TOTAL | REF. |
|---|---|---|---|
| 1 | UNIT LABOR AND MATERIAL COST FOR FOOTING FORMS | | WALKER P. 217 |

| MATERIAL @ 100 Ft² | Ft. BM | RATE | TOTAL |
|---|---|---|---|
| SIDES 2"X8" PLANKS | 133 | .05* | $6.65 |
| 2"X6" PLANKS | 100 | .047* | 4.70 |
| STAKES AND BRACES | 75 | .243 | 18.22 |
| NAILS | | | 1.20 |
| TOTAL @ 100 Ft² | | | 30.78 |
| TOTAL @ Ft² FOR MATERIALS | | | $0.31 |

TOTAL: $0.31

* BASED ON SIX USES OF LUMBER
2"X8" COST = $.327 PER Ft. BM. / 6 USES = .05
2"X6" COST = .285 PER FT. BM / 6 USES = .047

| LABOR @ 100 Ft² | HOURS | RATE | TOTAL |
|---|---|---|---|
| CARPENTER | 3.50 | 6.67 | $23.35 |
| LABOR | 1.75 | 4.50 | 7.88 |
| TOTAL @ 100 Ft² | | | 31.23 |
| TOTAL @ Ft² FOR LABOR | | | 0.31 |

TOTAL: $0.31

| NO | ITEM | TOTAL | REF. |
|---|---|---|---|
| 2 | REINFORCING BARS - HORIZONTAL No. 6 | | WALKER P. 321 |

TONS OF REBAR

LAP = 30 DIAMETERS = 30 X .750" = 22.5"/BAR
22.5" = 1.87 Ft.
BARS MUST LAP 1.87 Ft FOR EACH STANDARD 20 Ft. SECTION

THIS GIVES A % LAP OF 1.87 Ft / 20 Ft = 9%

644 LIN. Ft. X 3 BARS X 1.09 = 2106 LIN. Ft.
2106 LIN. Ft. REQUIRED / 20 Ft. STD. LENGTH = 105.3 BARS
MUST ORDER 106 BARS X 20 Ft = 2120 LIN. Ft.

2120 LIN. Ft. X 1.502 #/Ft = 3184 # / 2000 #/T. = 1.59 T

TOTAL: 1.59 T

| CONCRETE | BACK-UP SHEET | COMMERCIAL BLDG. | 3 |
|---|---|---|---|

| No | ITEM | TOTAL | REF. |
|---|---|---|---|
| 1 | REINFORCING BARS | | WALKER P. 324 |
| | LABOR PLACEMENT UNIT COST | | |

| TRADE | HOURS | RATE | TOTAL |
|---|---|---|---|
| STEEL WORKER | 16 | 7.67 | 122.72 |
| TOTAL COST @ TON | | | 122.72 |

TOTAL: 122.72

2

REINFOCING BARS - VERTICAL No. 4

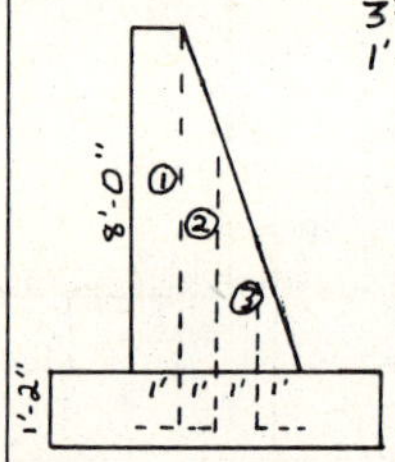


3" COVER = .25'
1'- LENGTH OF BENT PORTION

LENGTH OF BARS

① (8'-.25)+(1.17-.25)+1 = 9.67

② $\frac{8}{3}=\frac{X}{2}$ X = 5.33 Ft

(5.33-.25)+(1.17-.25)+1 = 7.00

③ $\frac{8}{3}=\frac{X}{1}$ X = 2.66 Ft

(2.66-.25)+(1.17-.25)+1 = 4.33

TOTAL LENGTH 21 Ft

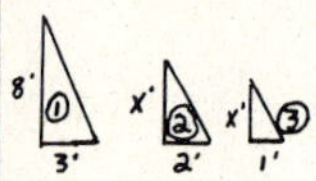


NUMBER OF BARS

644.0 Ft. ÷ 6 Ft O.C. = 108 SPACES
109 SETS OF BARS

① 9.67 (2) = 19.34 Ft, THEREFORE 54.5 BARS

② 7.00(2)+4.33 = 18.33 Ft, THEREFORE 54.5 BARS

③ 4.33 (4) = 17.32 Ft, THEREFORE 13.6 BARS

TOTAL 20' BARS 123 BARS

123 BARS X 20 Ft = 2460 LIN. Ft.

2460 LIN. Ft. X .668 = 1644# / 2000#/T = 0.82 T

TOTAL: 0.82 T

| CONCRETE | BACK-UP SHEET | COMMERCIAL BLDG. | 4 |
|---|---|---|---|

| No. | ITEM | TOTAL | REF. |
|---|---|---|---|
| 1 | LABOR UNIT COST FOR CONCRETE FOUNDATIONS POURED DIRECTLY | $2.01 | WALKER P. 363 |

| TRADE | HOURS | RATE | TOTAL |
|---|---|---|---|
| FORMAN | 0.08 | 6.67 | $0.53 |
| LABOR | 0.33 | 4.50 | 1.48 |
| @ C.Y. | | | $2.01 |

| No. | ITEM | TOTAL | REF. |
|---|---|---|---|
| 2 | LABOR UNIT COST FOR STRIPING FORMS | $0.07 | WALKER P. 217 |

| TRADE | HOURS | RATE | TOTAL |
|---|---|---|---|
| LABOR | 1.5 | $4.50 | $6.75 |
| @100 Ft. SQ. | | | 6.76 |
| @ Ft. SQ | | | $0.07 |

| No. | ITEM | TOTAL | REF. |
|---|---|---|---|
| 3 | FOUNDATION WALL - LINEAL FOOT CALCULATION | | B/U 1-1<br>S-1 A/S-2 |

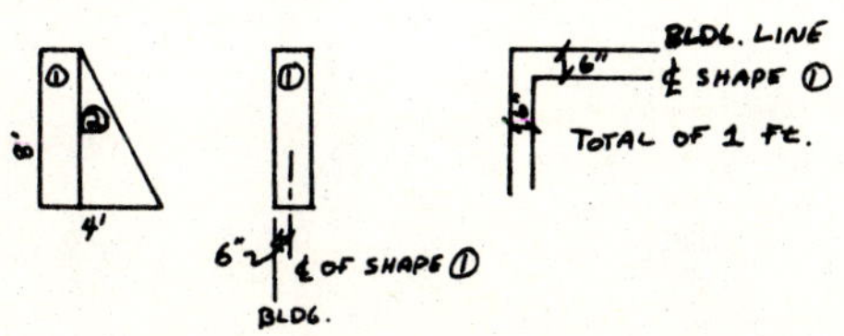


4 CORNERS x 1 Ft. PER. CORNER = 4.0 Ft

660 Ft. (LIN. Ft. BLDG. LINE) - 4.0 Ft = **656.0 Ft**

| CONCRETE | BACK-UP SHEET | COMMERCIAL BLDG. | 5 |
|---|---|---|---|

| NO. | ITEM | TOTAL | REF. |
|---|---|---|---|
| 1 | FOUNDATION WALL - LINEAL FOOT CALCULATION<br>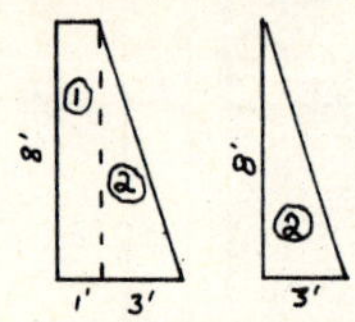<br>= 2/3 OF BASE<br>2/3 x 3Ft =<br>2Ft<br>THIS REQUIRES SAME ADJUSTMENT AS FOOTINGS SEE B/U 1-1 | 644.0Ft | S-1<br>A/S-2 |
| 2 | UNIT LABOR COST TO BUILD & ERECT FORMS | | WALKER P. 238 |

LABOR MAKING 100 SQ.Ft OF PANEL FORMS

| TRADE | HOURS | RATE | TOTAL |
|---|---|---|---|
| CARPENTER | 3.5 | $6.67 | $23.34 |
| COST PER 100 SQ Ft | | | $23.34 |
| COST PER SQ.Ft. | | | $0.23 |

LABOR ERECTING 100 SQ.Ft OF PANEL FORMS

| TRADE | HOURS | RATE | TOTAL |
|---|---|---|---|
| CARPENTER | 4.5 | $6.67 | $30.02 |
| LABOR | 2.25 | 4.50 | 10.13 |
| COST PER 100 SQ. Ft. | | | 40.15 |
| COST PER SQ.Ft | | | $0.40 |

BASED ON SIX REUSES OF PANELS TOTAL UNIT LABOR COST

($0.23 ÷ 6 USES) + $0.40 = $0.44 — TOTAL: $0.44 PER SQ Ft

| CONCRETE | BACK UP SHEET | COMMERCIAL BLDG | 6 |
|---|---|---|---|

| No. | ITEM | TOTAL | REF. |
|---|---|---|---|
| 1. | UNIT MATERIAL COST FOR WALL FORM<br>LUMBER COST – 1 2'x8' PANEL | | |

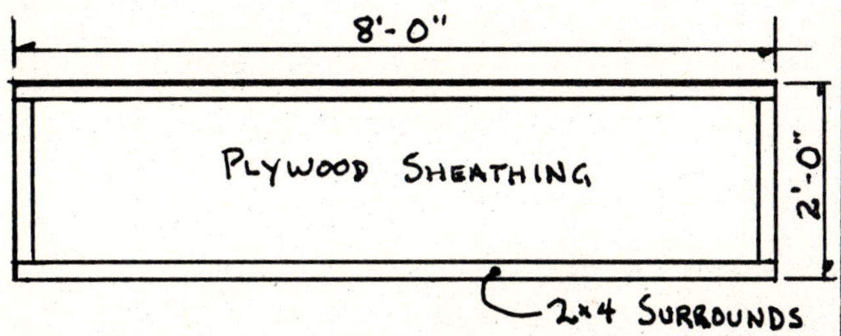


| ITEM | QUANTITY | RATE | TOTAL |
|---|---|---|---|
| 2x4- 20LF | 13.3 BF | $0.243 | $3.24 |
| PLYFORM | ½ PC | 19.34 | 9.67 |
| NAILS | — | L/S | 0.50 |
| WALES, 2-2x4-32" O.C. | 8.0 BF | 0.243 | 1.95 |
| COST PER PANEL | | | $15.36 |

REF.: WALKER P. 223

ASSUME 6 USES 15.36/6 $2.56

$2.56/16 S.F. $0.25/S.F.
SNAP TIES $0.30 EA – 2'0" O.C 0.08/S.F.
NAILS & EXPENDABLE BRACING 0.02/S.F.

REF.: WALKER P. 231

TOTAL UNIT PRICE FOR FORM PANELS — TOTAL: $0.25 PER S.F.

| CONCRETE | BACK-UP SHEET | COMMERCIAL BLDG. | 7 |
|---|---|---|---|

| No. | ITEM | TOTAL | REF. |
|---|---|---|---|
| 1 | UNIT LABOR COST FOR PLACING CONCRETE INTO FOUNDATION WALL FORMS USEING A CRANE AND BUCKET. | | WALKER P. 364 |

| TRADE | HOURS | RATE | TOTAL |
|---|---|---|---|
| LABOR | 0.40 | $4.50 | 1.8 |
| CRANE OP. | 0.05 | 9.60 | .48 |
| FOREMAN | 0.05 | 6.67 | .33 |
| | | | $2.61 |

TOTAL: $2.61

| No. | ITEM | TOTAL | REF. |
|---|---|---|---|
| 2 | UNIT CRANE RENTAL COST FOR FOUNDATION | | WALKER P. 364 |

QUANTITY PLACED PER DAY 150 TO 170 cu.yd.

481 cu.yd IN FND. WALL ÷ 160 cu.yd. PER DAY = 3 DAYS

| EQUIPMENT | DAYS | RATE | TOTAL |
|---|---|---|---|
| CRANE | 3 | $100.00 | $300.00 |
| @ 481 cu.yd. | | | $ 300.00 |
| @ cu.yd. | | | 0.62 |

TOTAL: $0.62

| No. | ITEM | TOTAL | REF. |
|---|---|---|---|
| 3 | UNIT LABOR COST FOR STRIPING FORMS | | WALKER P. 238 |

| TRADE | HOURS | RATE | TOTAL |
|---|---|---|---|
| LABOR | 2.25 | $4.50 | $10.13 |
| COST @ 100 SQ.Ft | | | 10.13 |
| COST @ SQ.Ft | | | $0.10 |

TOTAL: $0.10

| No. | ITEM | TOTAL | REF. |
|---|---|---|---|
| 4 | UNIT LABOR COST FOR LAYING WIRE MESH | | WALKER P. 327 |

| TRADE | HOURS | RATE | TOTAL |
|---|---|---|---|
| LABOR | 0.53 | $4.50 | $2.39 |
| COST @ 100 SQ.Ft | | | 2.39 |
| COST @ SQ.FT | | | $0.02 |

TOTAL: $0.02

| CONCRETE | BACK-UP SHEET | COMMERCIAL BLDG. | 8 |
|---|---|---|---|

| NO. | ITEM | TOTAL | REF. |
|---|---|---|---|
| 1. | AREA OF SLAB | | S-1 A S-2 |

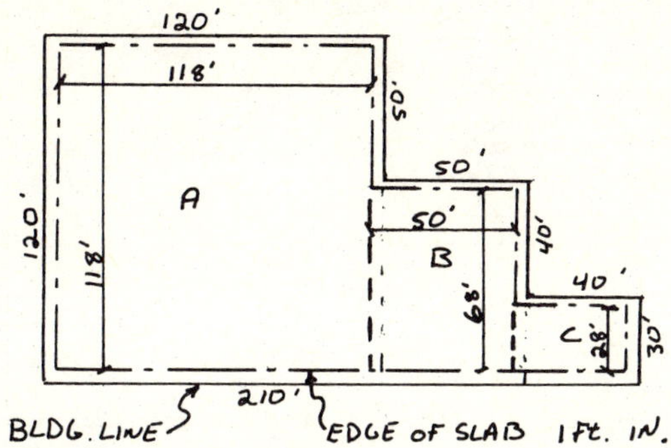


A. 118' X 118' = 13,924
B. 50' X 68' = 3,400
C 28' X 40' = 1,120
TOTAL 18,447 SQ.Ft.

TOTAL: 18,447 SQ. Ft.

| NO. | ITEM | TOTAL | REF. |
|---|---|---|---|
| 2 | FINE GRADING UNIT COST FOR SLAB | $0.07 | WALKER P. 164 |

| TRADE | HOURS | RATE | TOTAL |
|---|---|---|---|
| LABOR | 1.5 | $4.50 | $6.75 |
| @100 SQFt | | | 6.75 |
| @ SQ.Ft | | | $0.07 |

CONCRETE | BACK-UP SHEET | COMMERCIAL BLDG. | 9

| NO | ITEM | TOTAL | REF. |
|---|---|---|---|
| 1 | UNIT LABOR AND MATERIAL COST FOR FINISHING AND CURING | | WALKER 398 |

LABOR

| TRADE | HOURS | RATE | TOTAL |
|---|---|---|---|
| MASON OPERATOR | 8 | $7.58 | $60.64 |
| MASON | 16 | 7.58 | 121.28 |
| COST @ 4,000 SQ.FT. | | | 181.92 |
| COST @ SQ.FT. | | | $0.05 |

TOTAL: $0.05

MATERIAL (REF. WALKER P. 449)

| ITEM | NO | RATE | TOTAL |
|---|---|---|---|
| MACHINE | 8 | $7.00 | $40.00 |
| CURING COMPOUND | 12 GAL. | $1.75 | 21.00 |
| COST @ 4,000 SQ.FT. | | | 61.00 |
| COST @ SQ.FT. | | | 0.02 |

TOTAL: $0.02

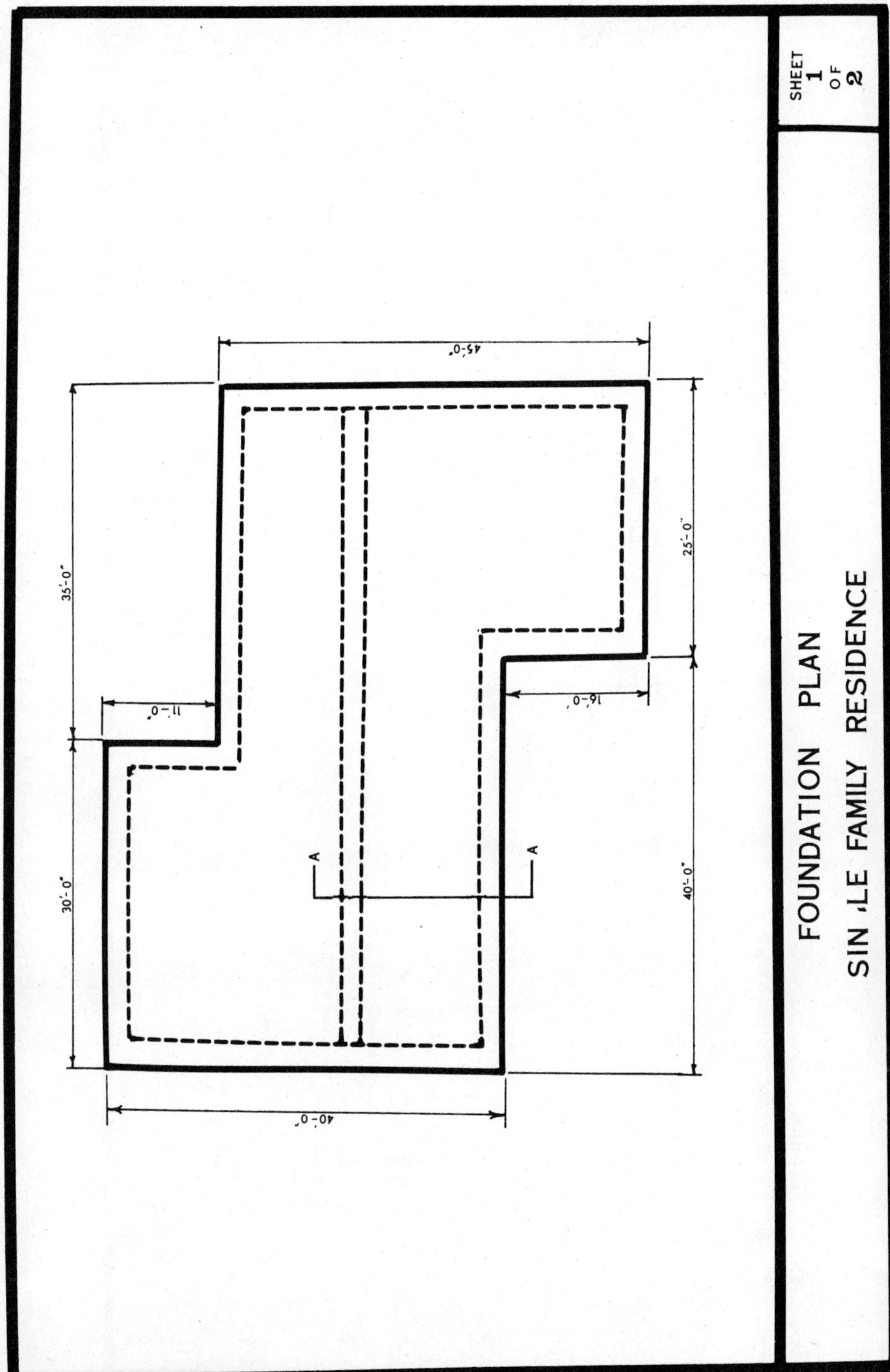
SHEET 1 OF 2
FOUNDATION PLAN
SIN ,LE FAMILY RESIDENCE
45'-0"
35'-0"
25'-0"
11'-0"
16'-0"
30'-0"
40'-0"
40'-0"
A
A

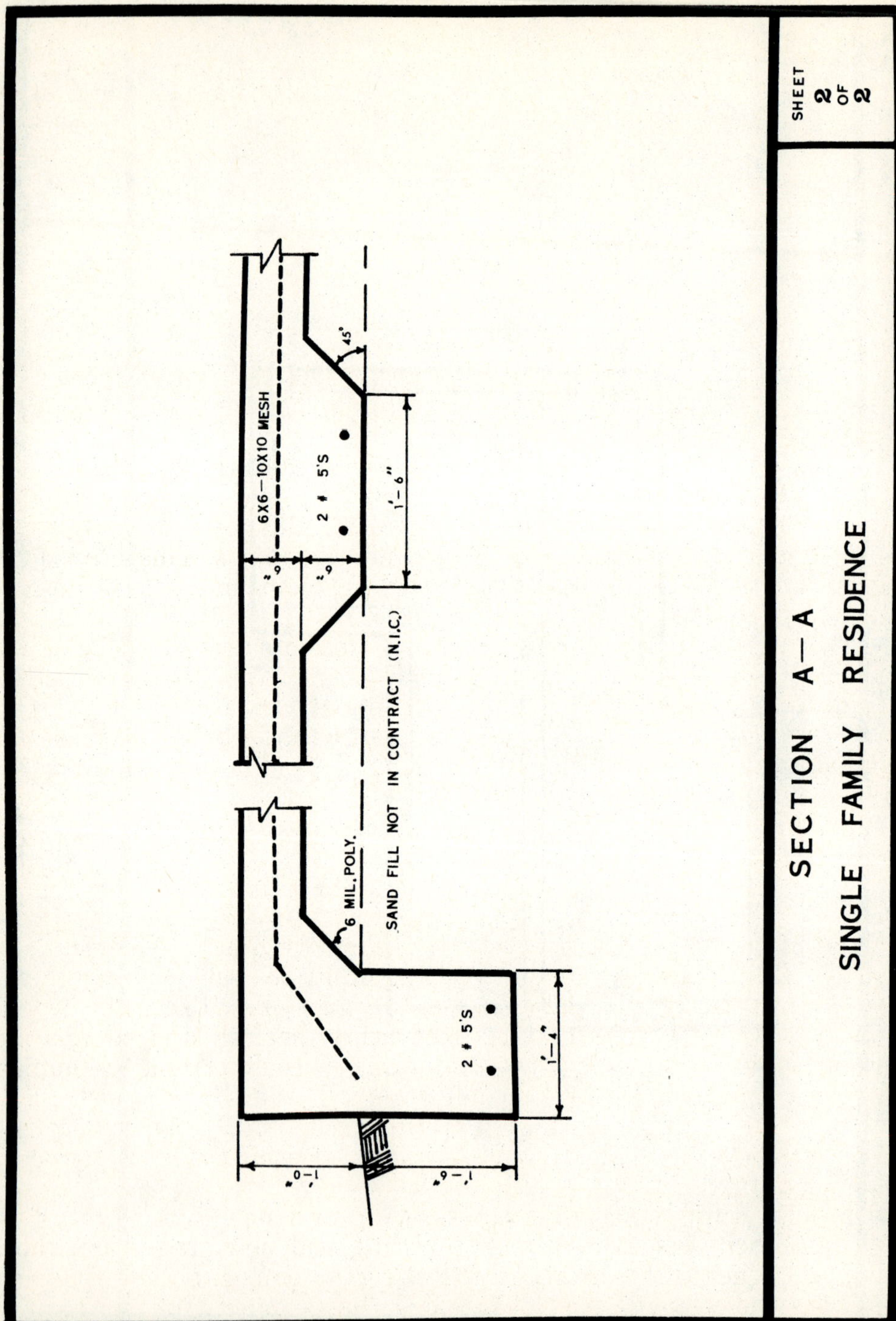
SHEET 2 OF 2
SECTION A — A
SINGLE FAMILY RESIDENCE
6X6—10X10 MESH
2 # 5'S
1'—6"
6"
6"
45°
SAND FILL NOT IN CONTRACT (N.I.C.)
6 MIL. POLY.
2 # 5'S
1'—4"
1'—0"
1'—6"

CHAPTER 6

# DIVISION 2 - SITEWORK

Sitework is the preparation of the site for the structure. It usually includes clearing of the site, earthwork and site drainage. These items are usually done before any other construction activities start. The last part of the sitework is preparation of the site for the owner. This includes pavements, walks and landscaping.

The estimator should familiarize himself with the site plan and should visit the site to check for existing structures, density of underbrush, soil conditions and site drainage. Most of this information is usually found on the site plan or specifications along with subsurface exploration reports. This site visit allows the estimator to check for any undetected problems and select the methods required to prepare the site for the structure.

The clearing of unwanted structures and trees are usually subcontracted to a specialist. The clearing of the site can range from removing multi-story buildings to small shacks. The estimator should try to receive a lump-sum cost for structure removal. The cost of clearing underbrush and trees can be figured by lump-sum or unit price subcontracts. Many subcontractors will provide bids of cost-per-acre requiring clearing and grubbing.

After the clearing of the site, the estimator must determine the cost of earthwork. Earthwork includes grading, excavation, backfill and compaction. When determining the cost of grading, the estimator must calculate the square foot area requiring grading. The procedure for estimating quantities for grading follows the same procedure as fine grading in Division 3 - Concrete. The major difference in site grading and fine grading is in the production rate. Site grading is usually done by a grader or small tractor providing a high production rate. The cost of excavation, backfill and compaction are based on the cubic yard, requiring the estimator to determine the cubic yards of soil to be handled. The following pages will show how to determine the quantity of soil requiring handling for the commercial building. Figure 34 is the site plan showing finished floor elevations, existing contours, new contours and specifications.

The specifications state that the existing and new contour lines do not include four inches of top soil. The existing and new grades are, therefore, four inches higher than shown on the site plan contours. The specifications also call for the four inches of top soil to be stripped and stockpiled to be

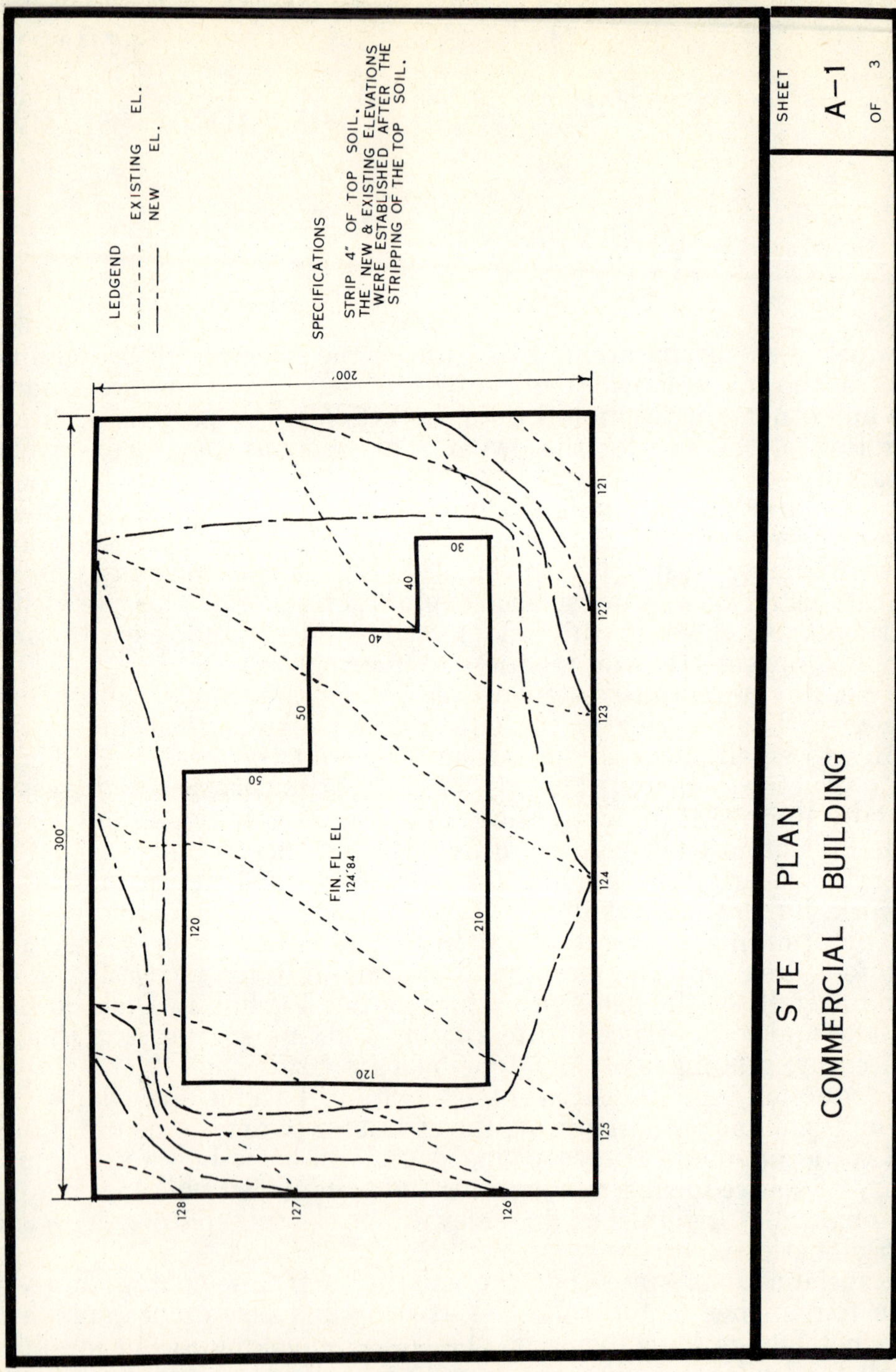
LEDGEND
EXISTING EL.
NEW EL.
SPECIFICATIONS
STRIP 4" OF TOP SOIL.
THE NEW & EXISTING ELEVATIONS
WERE ESTABLISHED AFTER THE
STRIPPING OF THE TOP SOIL.
200'
300'
FIN. FL. EL.
124.84
121
122
123
124
125
126
127
128
SHEET
A–1
OF 3
S TE PLAN
COMMERCIAL BUILDING

**FIGURE 33**

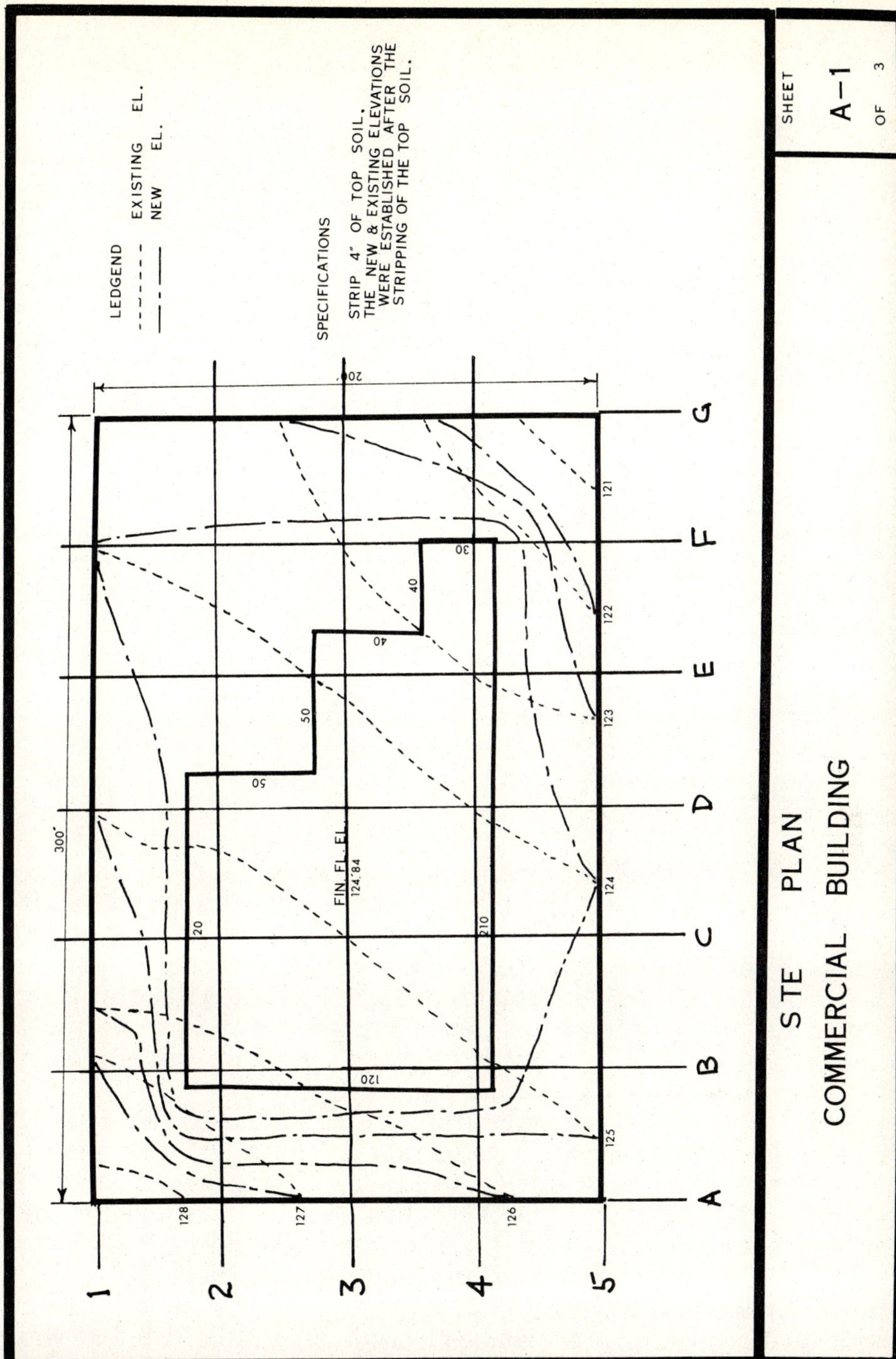
LEDGEND
- - - - - - EXISTING EL.
— - — NEW EL.
SPECIFICATIONS
STRIP 4" OF TOP SOIL.
THE NEW & EXISTING ELEVATIONS WERE ESTABLISHED AFTER THE STRIPPING OF THE TOP SOIL.
SHEET
A–1
OF 3
S TE PLAN
COMMERCIAL BUILDING
FIN. FL. EL. 124.84
300'
200'
1
2
3
4
5
A
B
C
D
E
F
G
128
127
126
125
124
123
122
121
120
20
210
50
50
40
40
30

FIGURE 34

spread over the site area not covered by construction. To estimate the quantity of material to be stripped, the area of the site is multiplied by the depth (length x width x depth, 300 x 200 x 0.33 = 19,800 Cubic Ft.). Earthwork is estimated in cubic yards. Therefore, the 19,800 cubic feet is divided by 27 cubic feet per cubic yard to give 733 cubic yards. To estimate the quantity of material that must be stockpiled to be spread over the site area not covered by construction, the estimator can multiply the surface area of the concrete structure (concrete take-off) by the depth of four inches, then subtract this amount from the 733 cubic yards computed above. [733 cubic yards - (19,600 area of concrete x .33 feet - depth of top soil ÷ 27 cubic feet per cubic yard) = 493 cubic yards].

The next activity is to determine the quantity of site excavation. The commercial building requires cuts at one end and fills at the other. Using the cross-sectional method, the estimator can determine if the site requires the purchasing of fill or the removal of excess material. The cross-sectional method is used when the plot plan shows existing and new contours. The system requires setting up a grid or dividing the area into squares or rectangles. The depths of cut or fill at the corners are computed. Then, the volume of each section is determined by averaging the depths and multiplying the average by the cross-sectional area of the section. Figure 34 shows how to set up a 50 foot grid system over the site plan.

After the grid system is set up, the estimator should figure the existing ground elevation of each grid line intersection by interpolation between the present contour lines. Then, the new ground elevations are figured for each intersection. This information should be transferred to a back-up sheet showing the existing elevations in the upper right quadrant and the new elevations in the upper left quadrant. When this is completed, the depth of cut or fill at each intersection can be computed by subtracting one elevation from the other and noting whether a cut or fill is required. This calculation should be made for each intersection and the results entered in the proper quadrant, using the lower left for cuts and the lower right for fills. After this information is computed for each intersection the estimator can determine the zero cut line. The zero cut line is where the excavation operation will change from cut to fill.

The volume of cut or fill is obtained by averaging the corner depth for each section, then multiplying by the area of that section. Figure 36 shows a back-up sheet with the grid lines, existing and new elevations, and the cuts or fills for the commercial building. Figures 37 thru 39 show a format for computing the total cut and fill required for the commercial building. The information for this activity is then transferred to the earthwork sheet.

The next earthwork activity is the excavation of the building foundation. To provide a safe working area, the excavation should be two feet wider than the footing and sloped at a 45 degree angle to the new grade elevation. Figure 40 shows the calculation of cut and fill for the commercial building.

QUANTITY SHEET

| PROJECT | ESTIMATOR | ESTIMATE NO. |
|---|---|---|
| LOCATION | EXTENSIONS | SHEET NO. OF |
| ARCHITECT ENGINEER | CHECKED | DATE |

CLASSIFICATION

| CODE | DESCRIPTION | S.F. (1) | C.Y. CUT (2) | C.Y. FILL (3) | TOTAL CY (4) | SKETCH & REF. (5) |
|---|---|---|---|---|---|---|
| | TOP SOIL | | | | | A-2 |
| | 300 X 200 X 0.33 | 60000 | 733 | — | — | |
| | 40400 SF X 0.33 | 40400 | — | 493 | — | |
| | 733 cy - 493 cy | — | — | — | 240 | REMOVED OFF SITE |
| | SITE EXCAVATION | | | | | |
| | B/U 1-1,2,3,4 | — | 839 | 358 | 481 | REMOVED OFF SITE |
| | BLDG. EXCAVATION | | | | | |
| | B/U 5-1, 5-2 | — | 3648 | 2943 | 705 | REMOVED OFF SITE |
| | TOTAL | | 5220 | 3794 | 1426 | |

| CODE | REF | DESCRIPTION | UNITS | L | M | LABOR | MATERIAL |
|---|---|---|---|---|---|---|---|
| 202.21 | | EXCAVATION | 5220 CY | — | — | — | — |
| 202.30 | | BACK FILL | 3794 CY | — | — | — | — |
| 202.50 | | COMPACTION | 3794 CY | — | — | — | — |
| 202.60 | | REMOVAL | 1426 CY | — | — | — | — |
| 202.40 | | SPREADING TOP SOIL | 493 CY | — | — | — | — |
| 202.11 | | SITE GRADING | 40400 SF | — | — | — | — |

QUANTITIES FOR SUBCONTRACT BIDS

**FIGURE 35**

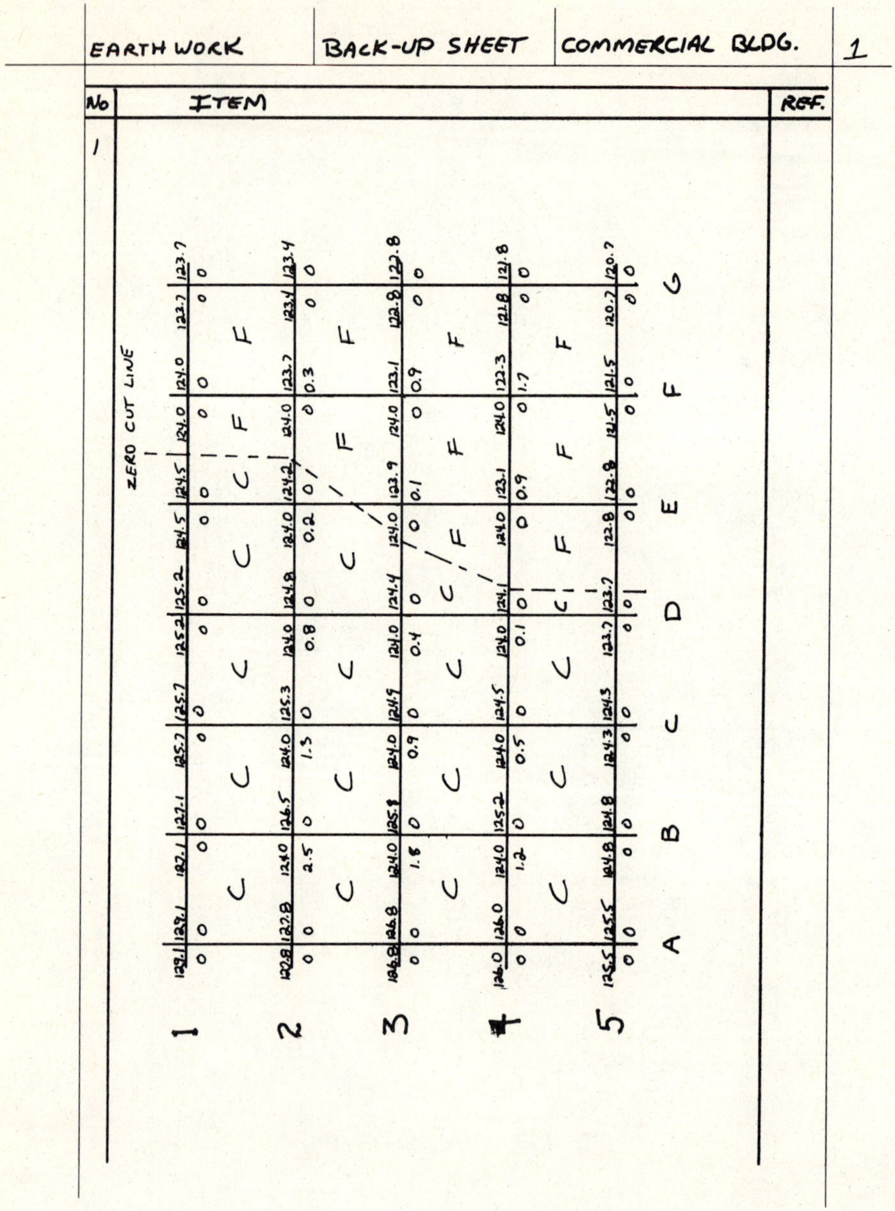
EARTH WORK
BACK-UP SHEET
COMMERCIAL BLDG.
1
No
ITEM
REF.
1
ZERO CUT LINE

**FIGURE 36**

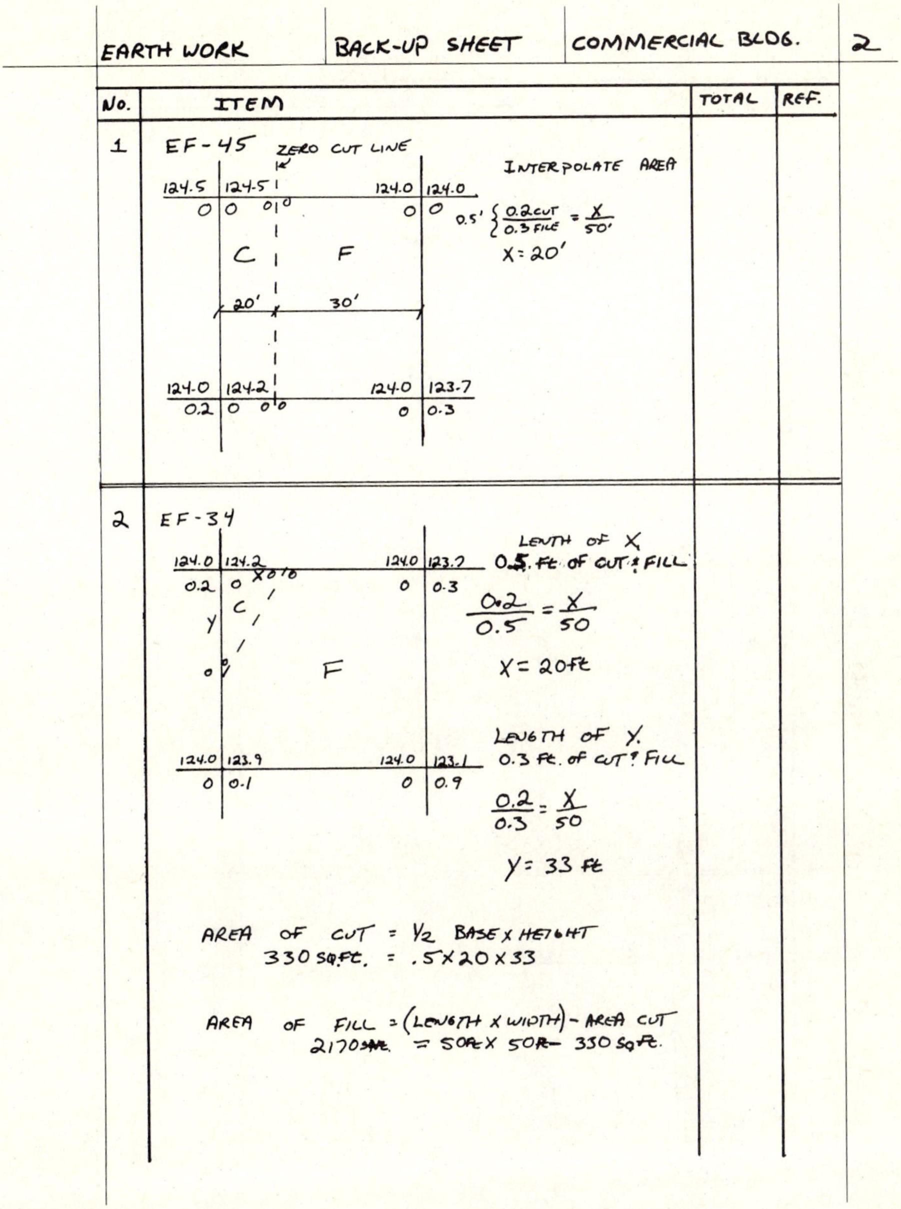
EARTH WORK
BACK-UP SHEET
COMMERCIAL BLDG.
2
No.
ITEM
TOTAL
REF.
1
EF-45
ZERO CUT LINE
124.5
124.5
124.0
124.0
0
0
0
0
0
0
C
F
20'
30'
124.0
124.2
124.0
123.7
0.2
0
0
0
0
0.3
INTERPOLATE AREA
0.5' { 0.2 CUT / 0.3 FILL = X / 50'
X = 20'
2
EF-34
124.0
124.2
124.0
123.7
0.2
0
X
0
0
0.3
C
Y
0
F
124.0
123.9
124.0
123.1
0
0.1
0
0.9
LENTH OF X
0.5 Ft. OF CUT & FILL
0.2 / 0.5 = X / 50
X = 20 Ft
LENGTH OF Y
0.3 Ft. OF CUT & FILL
0.2 / 0.3 = X / 50
Y = 33 Ft
AREA OF CUT = ½ BASE X HEIGHT
330 SQ.Ft. = .5 X 20 X 33
AREA OF FILL = (LENGTH X WIDTH) - AREA CUT
2170 SF. = 50 Ft X 50 Ft - 330 SQ.Ft.

**FIGURE 37**

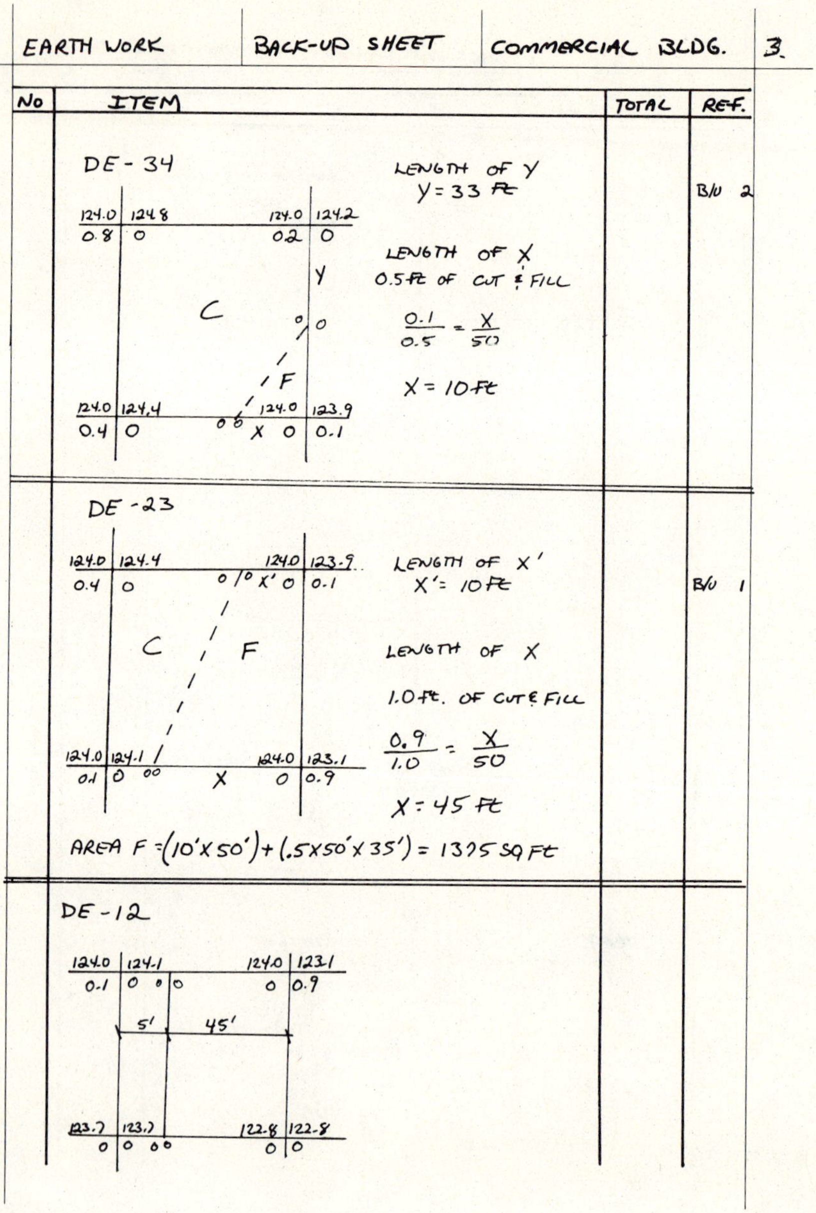
EARTH WORK
BACK-UP SHEET
COMMERCIAL BLDG.
3
No
ITEM
TOTAL
REF.
DE - 34
LENGTH OF Y
Y = 33 Ft
B/U 2
LENGTH OF X
0.5 Ft OF CUT & FILL
0.1 / 0.5 = X / 50
X = 10 Ft
DE - 23
LENGTH OF X'
X' = 10 Ft
B/U 1
LENGTH OF X
1.0 Ft. OF CUT & FILL
0.9 / 1.0 = X / 50
X = 45 Ft
AREA F = (10'X 50') + (.5X50'X 35') = 1375 SQ Ft
DE - 12

FIGURE 38

| EARTHWORK | BACK-UP SHEET | COMMERCIAL BLDG | 4 |
|---|---|---|---|

| NO. | ITEM | TOTAL | REF. |
|---|---|---|---|
| 1 | | | |

| GRID NO. | AVERAGE DEPTH | AREA | VOLUME | C | F |
|---|---|---|---|---|---|
| AB-45 | 0+0+0+2.5÷4=.6 | 2500 | 1500 | X | |
| AB-34 | 0+2.5+1.8+0÷4=1.1 | | 2750 | X | |
| AB-23 | 0+1.8+1.2+0÷4=.75 | | 1875 | X | |
| AB-12 | 0+0+0+1.2÷4=.3 | | 750 | X | |
| BC-45 | 0+0+2.5+1.3÷4=.95 | | 2375 | X | |
| BC-34 | 2.5+1.3+1.8+0.9÷4=1.63 | | 4062 | X | |
| BC-23 | 1.8+0.9+1.2+0.5÷4=1.1 | | 2750 | X | |
| BC-12 | 0+0+1.2+0.5÷4=.43 | | 1062 | X | |
| CD-45 | 0+0+1.3+0.8÷4=.53 | | 1325 | X | |
| CD-34 | 1.3+0.8+0.9+0.4÷4=.85 | | 2125 | X | |
| CD-23 | 0.9+0.4+0.5+0.1÷4=.48 | | 1188 | X | |
| FG-45 | 0+0+0.3+0÷4=.07 | | 188 | | X |
| FG-34 | 0.3+0.9+0+0÷4=.3 | | 750 | | X |
| FG-23 | 0.9+0+0+1.7÷4=.65 | | 1625 | | X |
| FG-12 | 1.7+0+0+0÷4=.43 | | 1063 | | X |
| EF-12 | 0.9+1.7+0+0÷4=.65 | | 1625 | | X |
| EF-23 | 0.1+0.9+0.9+1.7÷4=.9 | | 2250 | | X |
| DE-12 | 0+0+0+0.9÷4=.23 | ↓ | 653 | | X |

| GRID NO. | AVERAGE DEPTH | AREA | VOLUME | C | F |
|---|---|---|---|---|---|
| EF-45 | 0+0+0+0.3÷4=.08 | 1500 | 120 | | X |
| EF-45 | 0+0+0+0.2÷4=.05 | 1000 | 50 | X | |
| EF-34 | 0+0+0.3+0.9÷5=.24 | 2170 | 521 | | X |
| EF-34 | 0.2+0+0÷3=.06 | 330 | 20 | X | |
| DE-34 | 0.8+0.2+0.4÷5=.28 | 2335 | 654 | X | |
| DE-34 | 0.0+0.0+0.1÷3=.03 | 165 | 5 | | X |
| DE-23 | 0.1+0.9+0+0÷4=.25 | 1375 | 344 | | X |
| DE-23 | 0.1+0.4+0+0÷4=.13 | 1125 | 146 | X | |
| DE-12 | 0+0.9+0+0÷4=.23 | 2250 | 518 | | X |
| DE-12 | 0+0+0.1+0÷4=.02 | 250 | 5 | X | |

REF.: B/U - 1

| | TOTAL | |
|---|---|---|
| TOTAL CUT | 839 CY | |
| TOTAL FILL | 358 C.Y. | |

**FIGURE 39**

| EARTH WORK | BACK-UP SHEET | COMMERCIAL BULD. | 5 |
|---|---|---|---|

EXCUVATION FOR BUILDING FOUNDATION

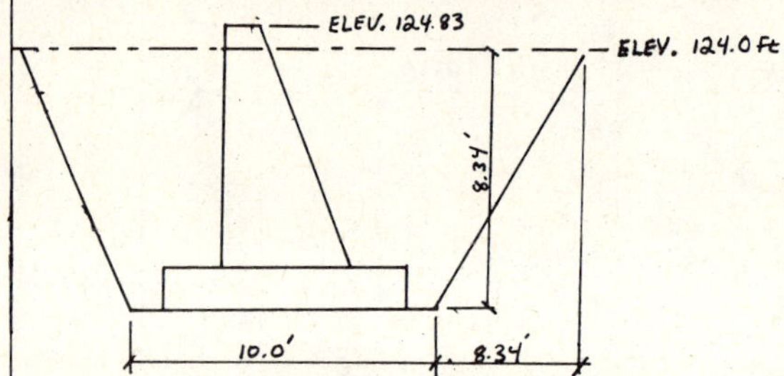


$\dfrac{644.0 \text{ LIN. FT. } ℄ \text{ FTG. } [(10.0' \times 8.34') + (8.34' \times 8.34')]}{27 \text{ C.F. per C.Y.}}$ = 3,648 CY CUT

BACK FILL FOR BUILDING FOUNDATION

705 CY OF CONCRETE IN BUILDING FOUNDATION

3,648 CY − 705 CY = 2,943 CY FILL

**FIGURE 40**

The complete quantity survey of earthwork is shown in Figures 35 through 40. The survey includes three activities: top soil, site excavation and building excavation. The operations required to complete these activities are excavation, back fill, compaction, removal of excess material, spreading of top soil, and site grading. Most general contractors do not have the capability to perform these operations; therefore, earthwork is usually subcontracted. Most earthwork subcontractors usually provide a separate unit cost bid for excavation, compaction, and material removal. For the estimator to establish the cost from these unit cost bids, he must make a quantity survey of each operation. If the earthwork subcontractor provides a lump sum bid, the estimator should still check the quantities of each operation to check the validity of the bid.

The cross-sectional method of determining excavation quantities provides an accurate quantity for the commercial building. A second method for determining quantities of excavation is the end area method. This method is used when cut and fill operations are to be performed over a long and narrow area such as roads and ditches. This method usually breaks the road or ditch into 100 foot stations. At these stations, the end areas are computed. To obtain the volume of cut and/or fill between each station, the end areas are averaged and then multiplied by the length of the station. This method is not as accurate as the cross-sectional method, but will provide the accuracy required for road and ditch estimating purposes.

The estimation of earthwork quantities will always have an element of approximation. The estimator must study soil reports to determine the soil characteristics. These soil characteristics will tell the estimator what swell factor to add to the bank cubic yards calculated. Also, these reports will show what type of material is to be excavated; sand, clay, rock, etc., and aid in the determination of equipment to be used. Most excavation for residential construction is not as complicated as the preceding example.

CHAPTER 7

# DIVISION 4 - MASONRY

Estimating the cost of masonry is not complex if the estimator understands the items which affect its cost. To establish the cost of a masonry wall, the estimator should know the class of workmanship, the type of mortar and joint, the type of building, and the method of bonding. With this knowledge, the estimator can calculate the quantities of brick, sand, mortor mix, scaffolding, ties, etc., and then develop a unit cost for laying the units.

The material unit cost for brick depends on the type of brick, mortar and bond required in the specifications. The labor unit cost depends on the style of the building, class of workmanship, mortar joint and bond required. The following step-by-step example of the commercial building's masonry estimate will show the estimator how these various items affect the cost.

The first step in preparing the masonry estimate is to set up the proper columns on the quantity sheet, Figure 41.

QUANTITY SHEET

PROJECT ESTIMATOR ESTIMATE NO.

LOCATION EXTENSIONS SHEET NO. OF

ARCHITECT ENGINEER CHECKED DATE

CLASSIFICATION

| | CODE | DESCRIPTION | 1<br>L. F | 2<br>S. F.<br>BRICK | 3<br>S. F.<br>8"CMU | 4<br>S. F.<br>12"CMU | 5<br>SKETCH<br>&<br>REF. | |
|---|---|---|---|---|---|---|---|---|
| 1 | | | | | | | | 1 |
| 2 | | | | | | | | 2 |
| 3 | | | | | | | | 3 |
| 4 | | | | | | | | 4 |
| 5 | | | | | | | | 5 |

**FIGURE 41**

Now the lineal feet of masonry wall must be calculated. This figure is used to calculate the quantity of brick, eight inch CMU's and joint reinforcement. There are two ways to calculate the number of masonry units required for the wall. The first method is called the course method. In this method the lineal feet of wall is divided by the length of the masonry unit. This gives the total number of units required for one course. Next, the height of the wall is divided by the height of the unit. This gives the number of courses of masonry unit. Then, the number of courses is multiplied by the number of units in one course. From this number, the number of masonry units in the windows and doors must be subtracted. This method takes considerable time to calculate the number of units in the windows and doors. The specifications require that the brick be bonded to the eight CMU's every six courses with metal ties. The course method, in this case, would provide an accurate estimate. If the specifications called for a full header every sixth course, the method would take too many calculations and would not provide accurate information.

The second method is called the square foot method. This method takes the square feet of wall, then multiplies it by a block or brick per square foot factor. This method allows the area of windows and doors to be multiplied by this factor to give the number of units to be deducted from the total. There are tables in *The Building Estimator's Reference Book* that show the number of brick and concrete block per square foot of wall area.

Using these tables, the quantities of brick and eight inch CMU's can be calculated for the commercial building shown on pages 80 and 81. The area of brick wall can be calculated by using the center line lineal foot of brick wall times the height. The adjustment of building line lineal foot to the center line of the brick is computed using the same method discussed in the Concrete Section (six exterior corners minus two interior corners leaves four offsetting corners, 4 corners x 0.33 ft. per corner = 1.32 feet of adjustment, 660.0 Lin. Ft. of bldg. line – 1.32 feet = 658.88 Lin. Ft. of brick center line). By using the center line measurement, the corner units are not counted twice. This center line distance is then multiplied by the height of the brick wall, plus two times the height of the parapet to give the area of brick wall. The area is then multiplied by the number of bricks per square foot to give the total number of bricks. The area of openings must then be calculated with the number of bricks in the opening substracted from the total. The Accounting Symbol of Parantheses ( ) is used to indicate a negative number. Figure 42 shows the masonry estimate with the calculation for the area of brick and block.

The next step is to develop the price block showing the operations required to lay the brick and block. The price block for the masonry estimate comes at the end of the estimate rather than after each item of work. Masons work from scaffolding, laying the units between knee and shoulder heights. The production rate for masonry work is constant, producing constant unit cost.

The first price block operation to consider is the unit labor and material cost for laying the brick. The area of wall requiring brick is multiplied by the number of bricks per square foot to give the number of bricks. This quantity should be the exact number of bricks. This exact quantity must be increased by a waste factor to take care of salmon brick, broken brick, bat, etc. There are several ways to incorporate this waste factor into the estimate. The first method is to deduct only large openings from the total square foot of wall. Another method is to deduct a percentage of the opening from the total,

QUANTITY SHEET

PROJECT ESTIMATOR ESTIMATE NO.
LOCATION EXTENSIONS SHEET NO. OF
ARCHITECT ENGINEER CHECKED DATE
CLASSIFICATION

| CODE | DESCRIPTION | 1 | 2<br>L.F. | 3<br>S.F. BRICK | 4<br>S.F. 8" CMU | 5<br>SKETCH & REF. |
|---|---|---|---|---|---|---|
| | BRICK | | | | | A-2 |
| | 658.65 X 12.38 | | 659 | 8154 | — | A |
| | | | | | | A-3 |
| | BLOCK 8" CMU | | | | | |
| | 654.66 X 9.38 | | 655 | — | 6140 | |
| | | | | | | |
| | LESS WINDOWS | | | | | |
| | 14- 4X5 | | 56 | (280) | (280) | |
| | 7- 3X7.33 | | 21 | (154) | (154) | |
| | 15- 5.33X3.67 | | 80 | (293) | (293) | |
| | LESS DOORS | | | | | |
| | 7-12.33X8.33 | | 86 | (718) | (718) | |
| | 4- 6X6.67 | | 24 | (160) | (160) | |
| | 10- 3X6.67 | | 30 | (200) | (200) | |
| | TOTAL | S.F. | | 6349 | 4335 | |

**FIGURE 42**

typically 25 percent. The method used in this manual is to add four percent to the brick per square foot. This increases the number of bricks by four percent to account for the waste.

The labor cost required to lay the brick depends on the standard of workmanship, mortar joint and style, type of bond, and type of building. The specification for the commercial building calls for common bond face brick, cement mortar, first class workmanship, and struck mortar joints. Using this information, the estimator must develop a labor unit cost. This labor cost should include laying the brick to prescribed specifications, transportation of brick at the job (labor hours), mixing of mortar, and placement of joint reinforcement. Most masonry contractors will develop this unit cost from historical data. If no historical data are available the estimator must count on reference books. *The Building Estimator's Reference Book* has several tables showing the labor production rate for various bonds, workmanship requirements, mortar and mortar joint styles. Using the production rates from *The Building Estimator's Reference Book*, a labor unit cost for laying brick and block is calculated in Figure 43. The unit material cost from brick is usually per thousand brick. For this reason, the unit labor cost for brick should be left

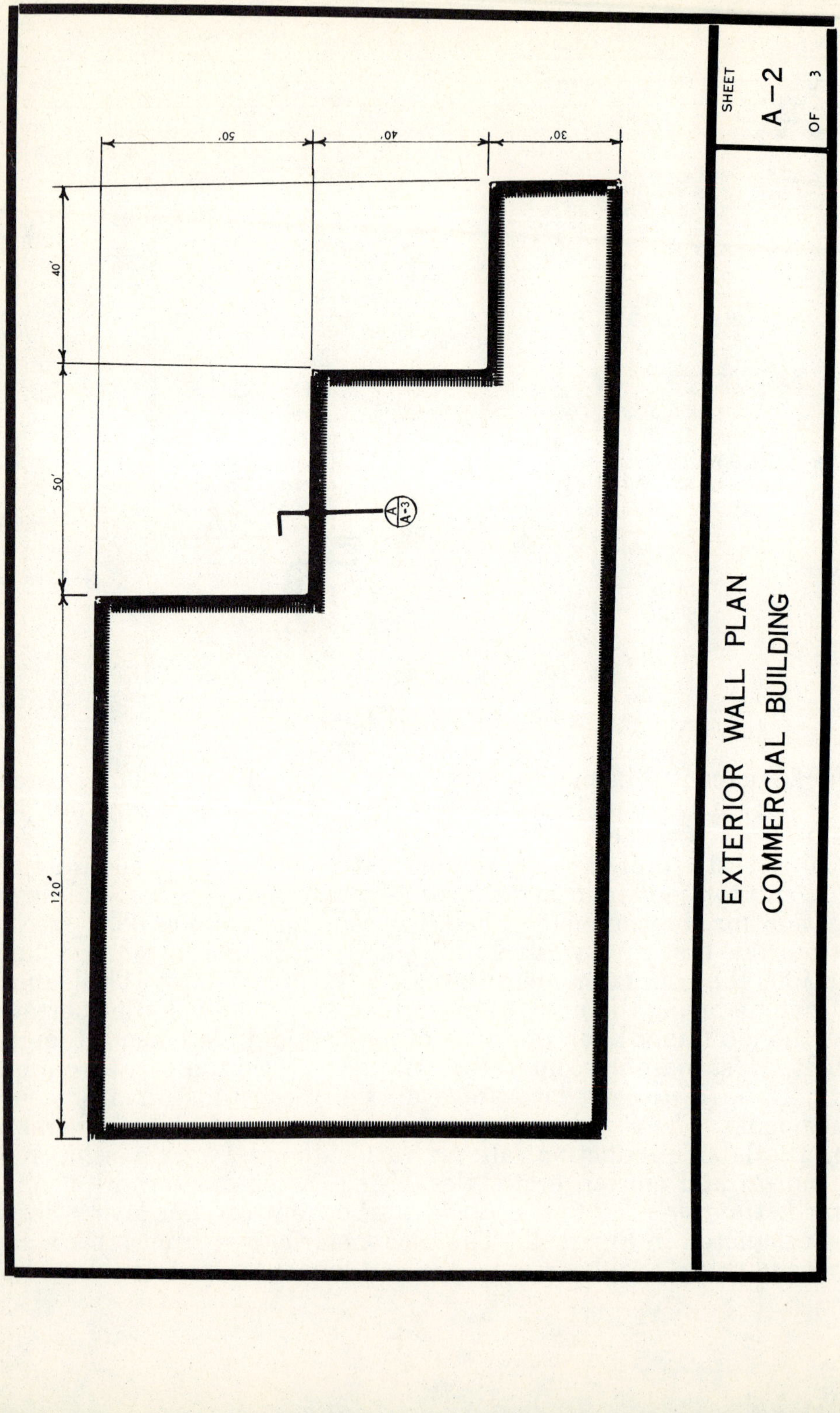
50'
40'
30'
40'
50'
120'
A
A-3
EXTERIOR WALL PLAN
COMMERCIAL BUILDING
SHEET
A-2
OF 3

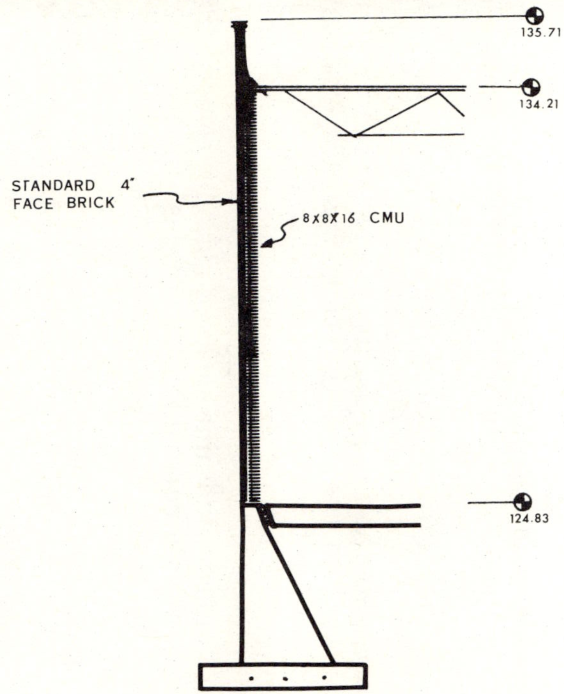


A/A-3 TYPICAL WALL SECTION

| WINDOW SCHEDULE ROUGH OPENING | |
|---|---|
| NO. | SIZE |
| 14 | 4'-0" x 5'-0" |
| 7 | 3'-0" x 7'-4" |
| 15 | 5'-4" x 3'-8" |

| DOOR SCHEDULE ROUGH OPENING | |
|---|---|
| NO | SIZE |
| 7 | 12'-4" x 8'-4" |
| 10 | 6'-0" x 6'-8" |
| 4 | 3'-0" x 6'-8" |

SPECIFICATIONS

BOND BRICK WITH METAL TIES.
3/8" MORTAR JOINTS
COMMON BOND
FACE SHELL BED 8" C.M.U.S
FIRST GRADE WORKMANSHIP
STRUCK JOINTS
CEMENT MORTAR
PRECAST LINTELS FOR BLOCK - 8" BEARING
3½"X 2½" METAL LINTELS FOR BRICK

WALL SECTION
COMMERCIAL BUILDING

SHEET
A - 3
OF 3

| MASONRY | BACK-UP SHEET | COMMERCIAL BLDG. | 1 |
|---|---|---|---|

| No. | ITEM | TOTAL | REF. |
|---|---|---|---|
| 1 | UNIT LABOR COST FOR BRICK | | |
| | | $227.91 @ 1000 | WALKER P.510 |
| | UNIT LABOR COST FOR 8" CMU | | |
| | | $0.61 | WALKER P. 547 |

UNIT LABOR COST FOR BRICK

| TRADE | HOURS | RATE | TOTAL |
|---|---|---|---|
| MASON | 21.5 | $ 7.67 | $164.91 |
| LABOR | 14 | 4.50 | 63.00 |
| COST @ 1000 BRICK | | | $227.91 |

UNIT LABOR COST FOR 8" CMU

| TRADE | HOURS | RATE | TOTAL |
|---|---|---|---|
| MASON | 5.0 | $7.67 | $38.35 |
| LABOR | 5.0 | $4.50 | 22.50 |
| COST@ 100 CMU | | | $60.85 |
| COST@ CMU | | | $ 0.61 |

**FIGURE 43**

in cost-per-thousand. The unit material cost for block is usually for one block, requiring the labor unit cost to be calculated for each block. The unit material cost for brick and block can be established by calling material suppliers.

The next operation is the estimate of the quantity of mortar. The unit labor cost for mixing mortar is included in the unit labor cost for laying the brick and block. The estimator must establish the quantity of mortar required to lay the masonry units. *The Building Estimator's Reference Book* provides several tables showing the quantity of mortar required for brick and block. The estimator must calculate the total quantity of mortar required and then break this quantity down into the quantity of mortar mix and sand. The quantity of mortar mix and sand is then converted to a unit material cost per cubic yard. Mortar proportion for average construction is one volume of masonry cement to three volumes of loose mortar sand. One cubic yard of mortar requires one cubic yard of sand and nine sacks of masonry cement. The mortar mix does not add any volume to the sand; it just fills the voids. The calculations of required quantities and material unit cost for the commercial building is shown in Figure 44.

The next item to be considered is scaffolding. Scaffolding is usually rented by the section and includes the required bracing and walkways. A good rule of thumb is to provide enough scaffolding to cover two sides of the building. Figure 45 shows the calculation of required sections for the commercial building. The estimator must call rental suppliers to get an updated cost per section.

Lintels for providing support for brick and block above an opening are the next items to be considered. The 3½" x 3½" steel lintels for the brick do not require a unit labor cost for placement. The precast lintels for the CMU's are

2 MATERIAL UNIT COST FOR MORTAR

SQUARE FOOT OF WALL
660 LIN. Ft X 12.38 Ft = 8171 SQ.Ft

(8171 SQ.Ft ÷ 100 SQ.Ft.) X 20.0 C.F = 1634 C.F.

1634 CF ÷ 27 CF/CY = 61 C.Y OF MORTOR — WALKER P. 489

REQUIRED QUANTITIES
SACKS OF MASONRY CEMENT 61 C.Y X 9 SACKS = 549 SACKS OF MIX

61 C.Y SAND.

MATERIAL UNIT COST PER CUBIC YARD

| ITEM | QUANTITY | RATE | TOTAL |
|---|---|---|---|
| MASONRY CEMENT | 9 SACKS | $3.25 | $29.25 |
| SAND | 1 C.Y. | 5.75 | 5.75 |
| COST PER C.Y | | | $35.00 |

$35.00 PER C.Y.

**FIGURE 44**

| MASONRY | BACK-UP SHEET | COMMERCIAL BLDG. | 2 |
|---|---|---|---|

| No | ITEM | TOTAL | REF. |
|---|---|---|---|

UNIT MATERIAL COST FOR SCAFFOLDING

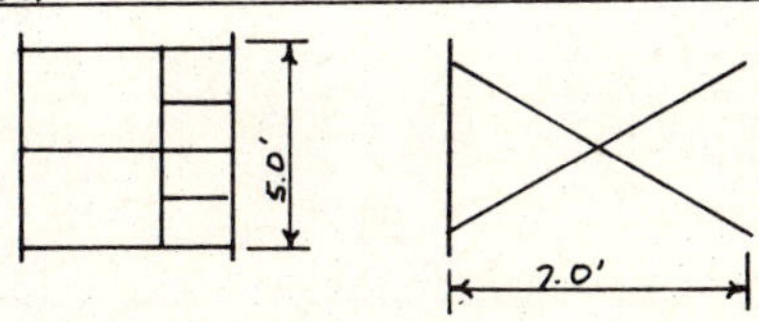


STANDARD SECTION 5.0' HIGH X 7.0' LONG.

LINEAL FOOT OF 2 SIDES OF BLDG. = 330 LIN.Ft.
330 LIN.FT ÷ 7.0' = 48 + 1 = 49 SECTION FIRST LEVEL

HEIGHT OF BUILDING = 10.71 Ft.
10.71 Ft ÷ 5.0' = 2 LEVELS

2 LEVELS X 49 SECTIONS = 98 SECTIONS — 100 SECTIONS

**FIGURE 45**

bulky and usually require a labor unit cost per lineal foot. This labor unit rate for the commercial building is set at $2.55 per lineal foot. The total lineal foot of openings must be increased by the bearing requirements.
The specifications require an eight inch bearing on each side of an opening. The commercial building has 57 openings. Therefore, the total lineal feet of lintels are computed as follows: 297 lineal feet of openings + (57 openings x 1.33 Ft.) = 371 lineal foot of lintels. The estimator should call suppliers to get the unit material cost for lintels. Joint reinforcement usually comes in two types; ladder and truss. Figure 49. The reinforcement is used to prevent cracks in the blockwork and/or to tie the brick & block together. The estimator must calculate the lineal feet of joint reinforcement. The metal ties to bond the brick to the block are usually estimated by the box. The estimator must use his judgement when estimating low cost items and develop a reasonable cost for these items.

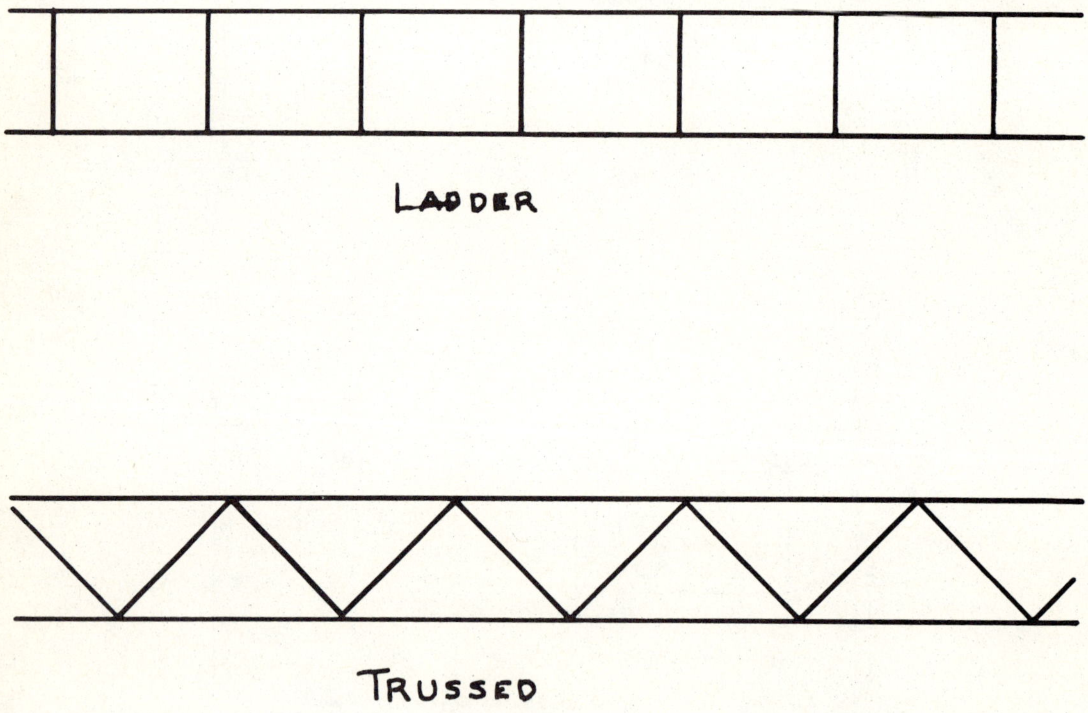


**FIGURE 46**

The last operation in the laying of the masonry wall is cleaning the brick. The brick is usually cleaned by masons and laborers using a weak solution of hydrochloric acid called Muriatic Acid. The estimator must determine the total area of brick wall to be cleaned. The area of brick wall was calculated when the quantity of bricks was calculated. The estimator must calculate a labor and material unit cost per square foot for cleaning.

Figures 47 thru 49 show the complete quantity take-off and back-up sheets for the commercial building. The price block shows the items which affect the cost of the masonry. This format provides a place for quantities and a price block to establish unit cost.

QUANTITY SHEET

| PROJECT | ESTIMATOR | ESTIMATE NO. |
|---|---|---|
| LOCATION | EXTENSIONS | SHEET NO. OF |
| ARCHITECT ENGINEER | CHECKED | DATE |

CLASSIFICATION

| CODE | DESCRIPTION | L.F. | S.F. BRICK | S.F. 8" CMU | SKETCH & REF. |
|---|---|---|---|---|---|
| | BRICK | | | | A-2 |
| | 658.65 X 12.38 | 659 | 8154 | — | A / A-3 |
| | BLOCK | | | | |
| | 654.66 X 9.38 | 655 | — | 6140 | |
| | LESS WINDOWS | | | | |
| | 14- 4X5 | 56 | (280) | (280) | |
| | 7- 3X7.33 | 21 | (154) | (154) | |
| | 15- 5.33X3.67 | 80 | (293) | (293) | |
| | LESS DOORS | | | | |
| | 7-12.33 X 8.33 | 86 | (718) | (718) | |
| | 4- 6.X6.67 | 24 | (160) | (160) | |
| | 10- 3 X 6.67 | 30 | (200) | (200) | |
| | | | 6349 | 4335 | |
| | TOTAL S.F. | | | | |

| REF | DESCRIPTION | UNITS | L | M | LABOR | MATERIAL |
|---|---|---|---|---|---|---|
| B/U 1-1 | BRICK-6349x6.65+4% | 43.9 M | 227.91 | 400.00 | 10007 | 17560 |
| B/U 1-1 | 8"CMU-4335x1.125+2% | 4974 EA | 0.61 | 0.62 | 3034 | 3084 |
| B/U 1-2 | MORTAR | 61 CY | — | 35.00 | — | 2135 |
| B/U 2-1 | SCAFFOLDING | 100 SEC | — | 2.00 | — | 200 |
| | JOINT REINFORCING | 1310 LF | — | 0.09 | — | 118 |
| | STEEL LINTELS | 335 LF | — | 0.36 | — | 120 |
| | PRE CAST LINTELS | 335 LF | 2.55 | 0.63 | 854 | 211 |
| | CLEANING | 6349 SF | 0.14 | 0.02 | 889 | 127 |
| | ACCESSORIES | — LS | — | — | — | 50 |
| | | | PAGE TOTAL → | | 14784 | 23605 |

**FIGURE 47**

MASONRY | BACK-UP SHEET | COMMERCIAL BLDG. | 1

| No. | ITEM | TOTAL | REF. |
|---|---|---|---|
| 1 | UNIT LABOR COST FOR BRICK | | |

| TRADE | HOURS | RATE | TOTAL |
|---|---|---|---|
| MASON | 21.5 | $7.67 | $164.91 |
| LABOR | 14 | 4.50 | 63.00 |
| COST @ 1000 BRICK | | | $227.91 |

TOTAL: $227.91 @ 1000 — REF.: WALKER P.510

UNIT LABOR COST FOR 8" CMU

| TRADE | HOURS | RATE | TOTAL |
|---|---|---|---|
| MASON | 5.0 | $7.67 | $38.35 |
| LABOR | 5.0 | $4.50 | 22.50 |
| COST @ 100 CMU | | | $60.85 |
| COST @ CMU | | | $0.61 |

TOTAL: $0.61 — REF.: WALKER P. 547

2 MATERIAL UNIT COST FOR MORTAR

SQUARE FOOT OF WALL

660 LIN. FT X 12.38 FT = 8171 SQ.FT

(8171 SQ.FT ÷ 100 SQ.FT.) X 20.0 C.F = 1634 C.F.

1634 CF ÷ 27 CF/CY = 61 C.Y OF MORTOR — REF.: WALKER P. 489

REQUIRED QUANTITIES

SACKS OF MASONRY CEMENT 61 C.Y X 9 SACKS = 549 SACKS OF MIX

61 C.Y SAND.

MATERIAL UNIT COST PER CUBIC YARD

| ITEM | QUANTITY | RATE | TOTAL |
|---|---|---|---|
| MASONRY CEMENT | 9 SACKS | $3.25 | $29.25 |
| SAND | 1 C.Y. | 5.75 | 5.75 |
| COST PER C.Y | | | $35.00 |

TOTAL: $35.00 PER C.Y.

**FIGURE 48**

| MASONRY | BACK-UP SHEET | COMMERCIAL BLDG. | 2 |
|---|---|---|---|

| NO | ITEM | TOTAL | REF. |
|---|---|---|---|
| | UNIT MATERIAL COST FOR SCAFFOLDING | | |

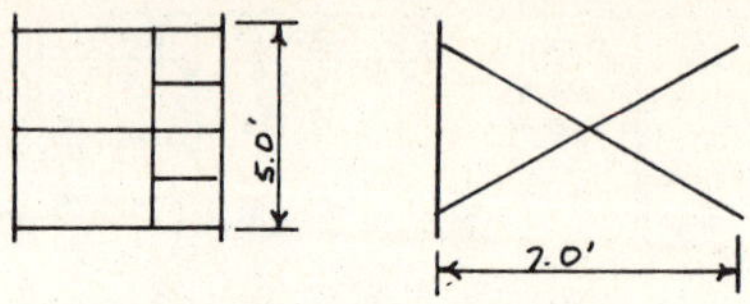


| NO | ITEM | TOTAL | REF. |
|---|---|---|---|
| | STANDARD SECTION 5.0' HIGH X 7.0' LONG. | | |
| | LINEAL FOOT OF 2 SIDES OF BLDG. = 330 LIN.FT. | | |
| | 330 LIN.FT ÷ 7.0' = 48 + 1 = 49 SECTION FIRST LEVEL | | |
| | HEIGHT OF BUILDING = 10.71 FT. | | |
| | 10.71 FT ÷ 5.0' = 2 LEVELS | | |
| | 2 LEVELS X 49 SECTIONS = 98 SECTIONS | 100 SECTIONS | |

**FIGURE 49**

CHAPTER 8

# DIVISION 6 - CARPENTRY

Carpentry is divided into three parts; rough, finish and architectural. Rough carpentry includes studs, plates, joist, roof sheeting, columns, beams, etc. In general, any wood that cannot be seen after the building is finished is considered rough. Finish carpentry includes items such as plywood paneling, sofits, finish siding, molding, facias, etc. Architectural woodwork includes custom cabinetwork, counter tops, doors, shelving, etc. The quantity take-off of these items requires the estimator to be skilled in carpentry plan reading and techniques. He must be able to distinguish the difference between rough and finish carpentry when looking at wall sections.

The first part of this section will deal with the quantity take-off of rough carpentry. Finish and architectural carpentry will be lumped together to make up the second part. This chapter will cover the format for the carpentry estimate, showing the carpentry activities and operations required in the price block. Because of the variables controlling the material and labor unit cost, they will be omitted. There are tables in *The Building Estimator's Reference Book* that show the number of brick and concrete block per square foot of wall area.

The format for the carpentry estimate follows the same basic data developed for concrete, earthwork and masonry. The five column accounting paper is set up with the proper heading and column titles. Each carpentry activity is listed in the description column, then the operations are listed in the price block. Figure 50 shows the heading and proper column titles for the carpentry take-off.

QUANTITY SHEET

PROJECT | ESTIMATOR | ESTIMATE NO.

LOCATION | EXTENSIONS | SHEET NO. OF

ARCHITECT ENGINEER | CHECKED | DATE

CLASSIFICATION

| | CODE | DESCRIPTION | 1<br>L. F. | 2<br>S. F. | 3<br>F. B. M. | 4<br>P. C. | 5<br>SKETCH & REF. | |
|---|---|---|---|---|---|---|---|---|
| 1 | | | | | | | | 1 |
| 2 | | | | | | | | 2 |
| 3 | | | | | | | | 3 |
| 4 | | | | | | | | 4 |

**FIGURE 50**

The lineal foot column is used to take off trusses, furring, facias, etc. This column is also used to calculate the footboard measure column. Table 5 shows the number of board feet per lineal foot of various sizes of lumber. The square foot column is used to take-off roof sheeting, wall sheeting, plywood, etc. The carpentry take-off, like the preceding take-offs, should follow a logical construction order. To illustrate the methods of surveying carpentry activities and how to follow a logical order, an example of a carpentry take-off for a single family residence will be studied.

The floor plan and wall sections for the residence are shown on the following two pages. This section will only deal with carpentry. Therefore, concrete foundation, block foundation, and wall activities will not be included. The first carpentry activity in the logical construction sequence is the exterior stud wall. This activity includes studs, top plates and bottom plate. The method for calculating the number of 2" x 4's" required for the plates is very simple. The total lineal feet of exterior wall is multiplied by the number of plates. The wall section for the residence calls for two top plates and one pressure treated bottom plate.

**Lineal Foot Table of Board Measure**
**Number of Feet of Lumber, B.M., Per Lineal Foot of any Size**

| | | |
|---|---|---|
| 2" x 4" = 0.667 | 4" x 4" = 1.333 | 8" x 14" = 9.333 |
| 2" x 6" = 1. | 4" x 6" = 2. | 8" x 16" = 10.667 |
| 2" x 8" = 1.333 | 4" x 8" = 2.667 | 10" x 10" = 8.333 |
| 2" x 10" = 1.667 | 4" x 10" = 3.333 | 10" x 12" = 10. |
| 2" x 12" = 2. | 4" x 12" = 4. | 10" x 14" = 11.667 |
| 2" x 14" = 2.333 | 4" x 14" = 4.667 | 10" x 16" = 13.333 |
| 2" x 16" = 2.667 | 4" x 16" = 5.333 | 10" x 18" = 15. |
| 2½" x 12" = 2.5 | 6" x 6" = 3. | 12" x 12" = 12. |
| 2½" x 14" = 2.917 | 6" x 8" = 4. | 12" x 14" = 14. |
| 2½" x 16" = 3.333 | 6" x 10" = 5. | 12" x 16" = 16. |
| 3" x 6" = 1.5 | 6" x 12" = 6. | 12" x 18" = 18. |
| 3" x 8" = 2. | 6" x 14" = 7. | 14" x 14" = 16.333 |
| 3" x 10" = 2.5 | 6" x 16" = 8. | 14" x 16" = 18.667 |
| 3" x 12" = 3. | 8" x 8" = 5.333 | 14" x 18" = 21. |
| 3" x 14" = 3.5 | 8" x 10" = 6.667 | 16" x 16" = 21.333 |
| 3" x 16" = 4. | 8" x 12" = 8. | 16" x 18" = 24. |

**TABLE 5**

The calculation of board feet of top and bottom plates is shown in the exterior wall section of Figure 51.

There are two methods for calculating the quantity of studs. The first method is to account for each stud in the exterior wall. The lineal feet of wall is divided by the spacing of studs to give the number of studs in the wall. This

| WINDOW SCHEDULE | | | |
|---|---|---|---|
| MARK | WIDTH | HEIGHT | NO |
| A | 3'—0" | 4'—2" | 12 |
| B | 3'—4" | 4'—6" | 4 |
| C | 2'—0" | 4'—2" | 2 |

| DOOR SCHEDULE | | |
|---|---|---|
| MARK | SIZE | NO |
| 1 | $3^0$ $7^0$ 1 3/4 | 2 |
| 2 | $2^8$ $6^8$ 1 3/8 | 3 |
| 3 | $3^0$ $6^8$ 1 3/8 | 1 |
| 4 | $5^0$ $6^8$ BYFOLD | 4 |

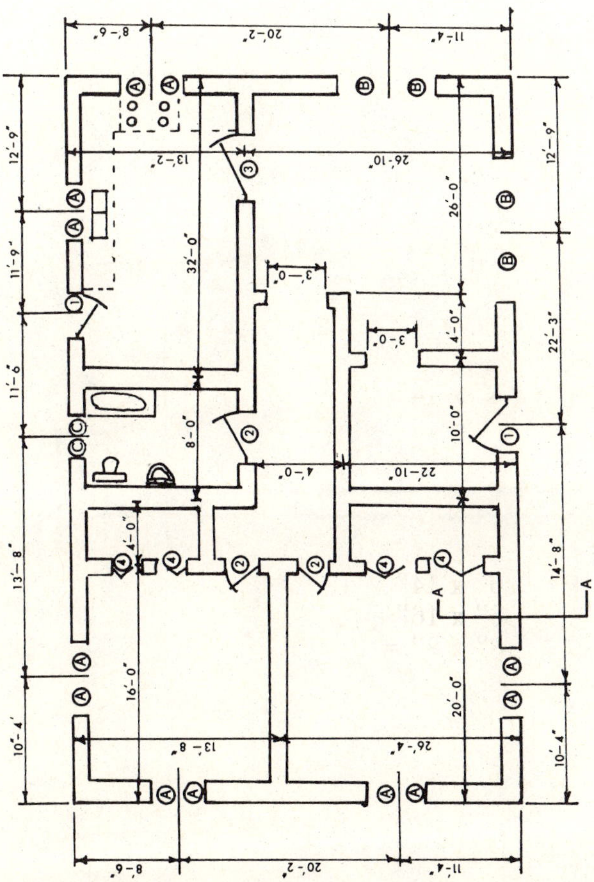

FLOOR PLAN

RESIDENCE

SHEET 1 OF 2

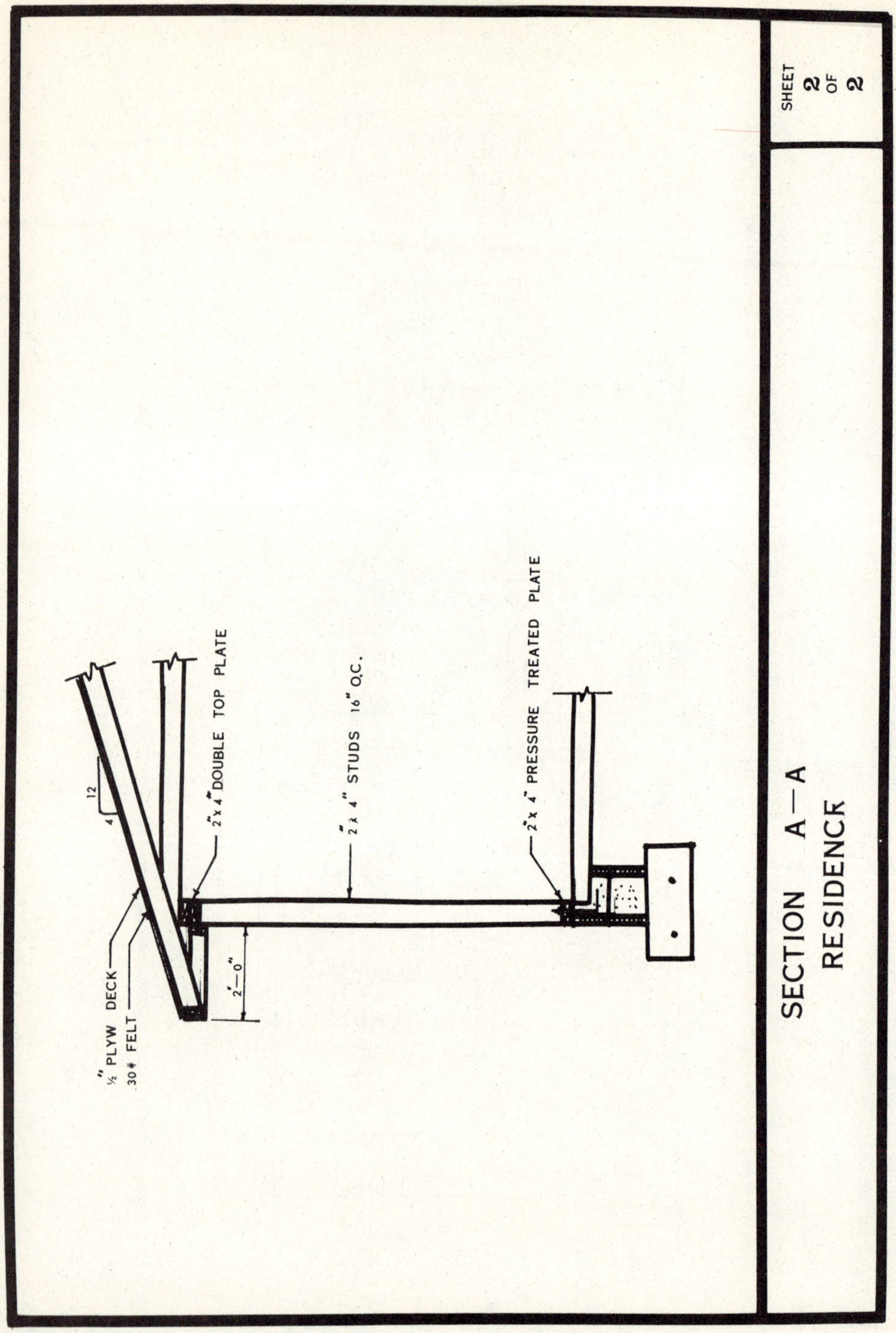
½" PLYW DECK
30# FELT
12
4
2"x4" DOUBLE TOP PLATE
2'—0"
2"x 4" STUDS 16" O.C.
2"x 4" PRESSURE TREATED PLATE
SECTION A—A
RESIDENCE
SHEET 2 OF 2

amount does not include double studs for window and door openings or corner studs. Codes require double studs on each side of openings exceeding three feet in width, and triple studs of each side of openings exceeding six feet in width. The code also requires not less than three studs to each exterior corner. Using the code requirements, the estimator can determine the number of studs required for the windows, doors, and corners. The second method uses a rule of thumb of one stud per lineal foot. This method does not include plates and is only good for 16 inch spacing. By using this rule of one stud per foot, the windows, doors and corners are accounted for. Figure 52 shows both methods of calculating the studs for the single-family residence. The window and door headers must now be calculated. The window and door openings six feet and under require the header bearing to be two inches on each side. The headers of six feet require four inches of bearing. The total lineal feet of window openings plus bearing will give the lineal feet of window and door opening. This number must be doubled to give the total lineal feet of double 2 x 8 headers.

The activity following the construction of the exterior wall is the placement of the roof truss. The quantity and pricing of trusses is usually based on lineal feet of bottom cord. The trusses for the residence are spaced twenty-four inches on center over the sixty foot length. To find the number of trusses, the length is divided by the spacing. This counts the spaces. Therefore, one truss must be added, times the width of the building. The calculation of the number of trusses in the residence is shown in the Roof Truss Section of Figure 53.

The roof sheating must now be attached to the roof trusses. To determine the square feet of roof sheating, the flat area of the roof plus overhang is multiplied by the roof slope factor. Table 6 shows the roof slope factor for various roof pitches.

**Roof Area Factors**

| **Pitch of Roof** | **Rise and Run** | **Multiply Flat Area by** | **Lin. Ft. of Hips or Valleys per Lin. Ft. of Common Run** |
|---|---|---|---|
| 1/12 | 2 in 12 | 1.014 | 1.424 |
| 1/8 | 3 in 12 | 1.031 | 1.436 |
| 1/6 | 4 in 12 | 1.054 | 1.453 |
| 5/24 | 5 in 12 | 1.083 | 1.474 |
| 1/4 | 6 in 12 | 1.118 | 1.500 |
| 7/24 | 7 in 12 | 1.158 | 1.530 |
| 1/3 | 8 in 12 | 1.202 | 1.564 |
| 3/8 | 9 in 12 | 1.250 | 1.600 |
| 5/12 | 10 in 12 | 1.302 | 1.612 |
| 11/24 | 11 in 12 | 1.357 | 1.685 |
| 1/2 | 12 in 12 | 1.413 | 1.732 |

**TABLE 6**

| CODE | DESCRIPTION | L.F | S.F. | FBM. | P.C. | SKETCH & REF. |
|---|---|---|---|---|---|---|
| | EXTERIOR WALL | | | | | |
| | | | | | | |
| | 2"x4" TOP PLATES | | | | | |
| | 200 LIN. Ft. X 2 PLATES | 400 | ——— | 287 | ——— | 0.667 B.F. |
| | | | | | | PER LIN. Ft. |
| | 2"x4" P.t. BOTTOM | | | | | |
| | PLATE | | | | | |
| | 200 LIN. Ft. X 1 PLATE | 200 | ——— | 134 | ——— | |

**FIGURE 51**

| CARPENTRY | BACK-UP SHEET | SINGLE FAMILY RESIDENCE | 1 |
|---|---|---|---|

| No | ITEM | TOTAL | REF. |
|---|---|---|---|
| 1 | QUANTITY OF STUDS IN EXTERIOR WALL | | |
| | STUD SIZE 2"X4"-8' | | |
| | 200 Ft. OF WALL ÷ 1.33' O.C. = 151 STUDS | | |
| | 2 OPENINGS REQUIRING 3 STUDS = 12 STUDS | | |
| | 9 OPENINGS REQUIRING 2 STUDS = 36 STUDS | | |
| | 4 CORNERS REQUIRING 3 STUDS = 12 STUDS | | |
| | 211 STUDS | 211 STUDS | |
| | SHORT CUT METHOD OF ONE STUD PER LINEAL FOOT 200 STUDS | | |

**FIGURE 52**

| CODE | DESCRIPTION | L.F | S.F. | FBM. | P.C. | SKETCH & REF. |
|---|---|---|---|---|---|---|
| | **EXTERIOR WALL** | | | | | |
| | 2"X4" TOP PLATES | | | | | |
| | 200 LIN.Ft X 2 PLATES | 400 | — | 287 | — | 0.667 B.F. |
| | | | | | | PER LIN. Ft. |
| | 2"X4" P.t. BOTTOM | | | | | |
| | PLATE | | | | | |
| | 200 LIN.Ft X 1 PLATE | 200 | — | 134 | — | |
| | 2"X4" STUDS | | | | | B/U 1-1 |
| | 211 STUDS X 8Ft X 0.667 | — | — | 1126 | 211 STUDS | |
| | 2" X 8" HEADERS X(2) | 126 | | 168 | | 1.33 BF |
| | | | | | | PER LIN. Ft |
| | **ROOF TRUSS** | | | | | WINDOW & DOOR SCHEDULE |
| | (60Ft ÷ 2' o.c.) X 40Ft | 1240 | — | — | 31 TRUSSES | |

FIGURE 53

The quantities of interior wall studs and plates are the next items in the construction sequence. The same methods used to calculate the material in the exterior wall will work for the interior partitions. When using the rule of thumb method of one stud per foot, the total lineal feet of interior partition wall should include the door openings. Figure 54 shows the plywood roof deck and interior wall calculations.

All of the activities for rough carpentry of the single-family residence are complete. The price block should now be developed showing the operations of each activity. The quantities for each activity are exact quantities. These quantities should be increased between three and five percent to account for waste and poor materials. The increase in quantities should be made when they are transferred to the price block. The price block activities which have the same unit labor and material cost should be lumped together. The first activity to be considered is the 2" x 4" pressure treated bottom plate. The board foot measure of bottom plate should be increased by the waste factor and inserted into the price block. The exterior and the interior 2" x 4" studs and top plates are the next activities to be transferred to the price block. These activities can be lumped together because they require the same unit labor and material cost. The board foot measures of these items are summed and then increased by the waste factor. Following the wall operation, the 2" x 8" headers are increased by the waste factor and included in the price block.

The roof truss activity does not require a waste factor because they are fabricated off the job. The net lineal feet of bottom chord is inserted in the price block. The last activity in the rough carpentry section is the plywood roof deck. This activity includes two operations; placement of the plywood and placement of one layer of 30 pound felt. Figure 54 shows the complete rough carpentry estimate through the units column of the price block. The unit material column can be completed by calling material suppliers for unit cost. The unit labor cost can be computed from historical data or by using reference books for production rates. The items shown in the rough carpentry estimate of the single-family residence are a small portion of what might be encountered when estimating a complex wood structure. If the estimator uses the carpentry format and follows the logical sequence of rough carpentry activities, he should not have any problems. The estimator can use the rough carpentry section of Appendix A to provide a guide to rough carpentry activities and the units required in the take-off.

The finish and architectural carpentry estimate should follow the same format. The column headings should be the same as in rough carpentry and the estimator should follow the logical order of finish carpentry activities. The first finish carpentry activity should be the exterior finish siding and facing materials. The exterior siding should be taken-off by the square foot with no deductions for openings. Next, the square feet of soffits and lineal feet of exterior trim must be calculated.

The second finish carpentry activity is the interior finish and facing materials. These items should be taken-off by the square foot and include plywood paneling, boards, etc. When estimating these quantities, no deductions should be made for openings in the wall area. The estimator can use the lineal foot dimension used for interior and exterior wall studs and plates times the wall height to calculate the area of interior wall surfaces. The area of ceiling is calculated by using the building dimension instead of room by room. These short-cut methods provide the accuracy required for the survey.

Architectural carpentry is the last carpentry item to be discussed. The activities included in architectural carpentry are custom cabinet work, counter tops, door, molding, trim, stairs and custom paneling. A lump sum cost for custom cabinet work and counter tops is frequently provided. The unit cost for doors is usually based on the number of doors. Molding, trim and custom paneling should be taken-off room by room using the proper units shown in Appendix A.

In commercial work, very little of the finish or architectural carpentry is done at the job. The interior finish woodwork is usually produced at mills. This millwork includes custom desks, cabinets, railing, counters, etc. In this case the estimator will receive a subcontract bid for millwork. In residential finish carpentry a quantity survey and cost estimate must be prepared by the estimator. Typically finish carpentry is a bid item in residential work.

PRACTICAL
FORM 516

QUANTITY SHEET

| PROJECT | ESTIMATOR | ESTIMATE NO. |
|---|---|---|
| LOCATION | EXTENSIONS | SHEET NO. |
| ARCHITECT ENGINEER | CHECKED | DATE |
| CLASSIFICATION | | |

| CODE | DESCRIPTION | 1 L.F | 2 S.F. | 3 FBM. | 4 P.C. | 5 SKETCH & REF. |
|---|---|---|---|---|---|---|
| | EXTERIOR WALL | | | | | |
| | 2"X4" TOP PLATES | | | | | |
| | 200 LIN. FT X 2 PLATES | 400 | — | 287 | — | 0.667 B.F. PER LIN. FT. |
| | 2"X4" P.t. BOTTOM PLATE | | | | | |
| | 200 LIN. FT X 1 PLATE | 200 | — | 134 | — | |
| | 2"X4" STUDS | | | | | B/U 1-1 |
| | 211 STUDS X 8FT X 0.667 | — | — | 1126 | 211 STUDS | |
| | 2"X8" HEADERS X (2) | 126 | | 168 | | 1.33 BF PER LIN. FT WINDOW & DOOR SCHEDULE |
| | ROOF TRUSS | | | | | |
| | (60FT ÷ 2' O.C.) X 40FT | 1240 | — | — | 31 TRUSSES | |
| | PLYWOOD ROOF DECK | | | | | |
| | 64FT X 44FT X 1.054 | — | 2968 | — | 93 SHEETS | 2 FOOT OVERHANG |
| | 1/2" PLYWOOD | | | | | |
| | INTERIOR WALL | | | | | |
| | 2"X4" TOP PLATE | | | | | |
| | 176 LIN. FT X 2 PLATES | 352 | — | 235 | — | |
| | 2"X4" Pt PLATE | 176 | — | 117 | — | |
| | 2"X4" STUDS | 176 | — | 939 | 176 EA. | |

| REF. | DESCRIPTION | UNITS | L | M | LABOR | MATERIAL |
|---|---|---|---|---|---|---|
| | 2"X4" pt. PLATE + 3% WAST | 258 BFM | x | x | x | x |
| | 2"X4" STUDS, PLATES 3% WAST | 2665 BFM | x | x | x | x |
| | 2"X8" HEADERS 3% WAST | 173 BFM | x | x | x | x |
| | ROOF TRUSS | 1140 LF. | x | x | x | x |
| | PLYWOOD ROOF DECK | 93 EA | x | x | x | x |
| | 30# FELT | 3146 SF | x | x | x | x |
| | | PAGE TOTAL → | | | x | x |

**FIGURE 54**

| CARPENTRY | BACK-UP SHEET | SINGLE FAMILY RESIDENCE | 1 |
|---|---|---|---|

| No | ITEM | TOTAL | REF. |
|---|---|---|---|
| 1 | QUANTITY OF STUDS IN EXTERIOR WALL | | |
| | STUD SIZE 2"X4"-8' | | |
| | 200Ft. OF WALL ÷ 1.33' O.C. = 151 STUDS | | |
| | 2 OPENINGS REQUIRING 3 STUDS = 12 STUDS | | |
| | 9 OPENINGS REQUIRING 2 STUDS = 36 STUDS | | |
| | 4 CORNERS REQUIRING 3 STUDS = 12 STUDS | | |
| | 211 STUDS | 211 STUDS | |
| | SHORT CUT METHOD OF ONE STUD PER LINEAL FOOT 200 STUDS | | |

**FIGURE 55**

CHAPTER 9

# BRINGING THE ESTIMATE TOGETHER

The final step in determining the cost of a structure is to bring all the various cost items together on summary sheets. The summary sheets should be set up in two parts; direct job cost and indirect job cost using the estimate summary sheets.

Direct job costs are the costs which can be attributed to a special construction activity. This includes the cost of activities calculated by the estimator; earthwork, concrete, masonry, carpentry, etc., and subcontractors bids; paint, tile, plumbing, electrical, HVAC, ect. Many subcontractors will provide a unit price bid for their specialty area. This requires the estimator to determine the quantity of materials before developing the cost. In many cases, the estimator will have already calculated these quantities. For example, the unit price for carpet is based on the square yard of floor area. The floor area is usually calculated in the concrete section. Another example: paint covers the same area as drywall in division 9 of the UCI code. The estimator can use Appendix A as a guide in determining which items are to be subcontracted and items for which he must prepare a detailed estimate.

Indirect costs are items which affect the cost of the structure but cannot be attributed to a specific construction activity. Division one - General Requirements of Appendix A provides a list of indirect costs. Many of these indirect cost items require the estimator to determine the duration of the job. From the job duration, the estimator can determine the cost of superintendents, temporary field office, temporary utilities and sanitary facilities, etc.

Overhead and Profit must include an allowance for a portion of the home office expenses as well as the costs of running the job office. The allowance for home office expense must include a share of salaries for management personnel, the costs of estimating and scheduling as well as rent, insurance and office utilities. These costs must be allocated to each job on the basis of anticipated business for the year as well as on historical data, showing these costs in past years. Profit is the object of every building operation. How much profit to add to a job is a question each builder must answer for each job. Some of the factors that should be considered before adding profit to the estimate are competition, local conditions and how badly the job is wanted.

The interest on the construction loan is another indirect cost item overlooked by some estimators. With the high interest rates of today, this is becoming a very important item. Most contractors usually set up accounts with material suppliers which become due on the tenth of each month and pay their employees at the end of each week. In commercial work the contractor will receive monthly draws from the owner. If he receives the draw before the tenth of the month, he will only need to cover his weekly labor cost. If he has predicted his cash flow accurately, he may not be required to borrow money. In residential work, the contractor typically receives money in three phases. The first phase is after the completion of the foundation. The second phase is after drying in the buildings. The final phase is at closing. By receiving funds in this way he can get by without borrowing the full construction amount for the full duration of the job. A good rule of thumb for computing the cost of construction financing is that the contractor will be required to borrow enough money to cover the cost of the building for one third of the construction period.

Taxes are the next items that must be included as an indirect cost. The employer's portion of Federal Income Tax, State and Federal Unemployment and Workman's Compensation are considered labor burden and is usually 20% of total labor cost. The total material cost should be increased by the State Sales Tax or Use Tax.

The last indirect cost items for commercial work are the cost of bonds and permits. As discussed in the Construction Review Section, the percentage of bond cost decreases as the cost of the structure increases. A good rule of thumb when estimating the cost of bond is that the bond will usually cost one percent of total job cost. In most cases, bond are not required in residential work and are not included in the summary sheet. The cost for the construction permit depends on the area of construction and is usually based on a certain cost per thousand dollars of job cost. The cost for permits can be established by a call to the local governing office.

Figures 56 and 57 show a typical summary sheet for a single-family residence. The first page of the summary sheets is for the direct job cost. The code column is used for the UCI's cost code which is used to update historical cost and check actual cost overruns. The description column is used to describe the cost item such as concrete, earthwork, masonry, electrical, plumbing, etc. These cost items should be listed in their logical construction order; The UCI code will help here. The units column is used to show the total units of each cost item. The labor and materials columns are used for the total labor or material cost. The subcontract and name columns are used to insert the subcontractor's bid and name. The second page of the summary sheet is for indirect job cost. The column heading will be used for the same information as direct job cost. Again you should notice that a small x is placed in all columns which require calculations and a line drawn through columns not to be used.

## SUMMARY OF ESTIMATE

BUILDING   LOCATION   PAGE   OF

ARCHITECT   OWNER   BID DATE/TIME

CUBICAL CONTENTS   NO. OF STORIES   COST PER CUBIC FOOT   ESTIMATOR

FLOOR, AREA, SQUARE FEET   COST PER SQUARE FOOT   CHECKER

| CODE | DESCRIPTION | UNITS | LABOR | MATERIAL | SUB. | NAME |
|---|---|---|---|---|---|---|
| 201.0 | CLEAR SITE | L.S | — | — | x | x |
| 202.2 | EARTH WORK | CY | x | x | — | — |
| 300.0 | CONCRETE | T.O. | x | x | — | — |
| 400.0 | MASONRY | T.O. | x | x | — | — |
| 500.0 | METALS | LB. | — | — | x | x |
| 601.0 | ROUGH CARPENTRY | T.O. | x | x | — | — |
| 602.0 | FINISH CARPENTRY | T.O. | x | x | — | — |
| 603.0 | ARCHITECTURAL WOOD | T.O. | x | x | — | — |
| 707.0 | SHINGLE ROOFING | SQ. | — | — | x | x |
| 708.0 | FLASHING | S.F. | — | — | x | x |
| 801.1 | DOORS | T.O. | x | x | — | — |
| 805.0 | WINDOWS | T.O. | x | x | — | — |
| 810.0 | HARDWARE | T.O. | x | x | — | — |
| 902.0 | GYPSUM DRYWALL | S.F. | — | — | x | x |
| 903.1 | CERAMIC TILE | S.F. | — | — | x | x |
| 908.2 | VINYL FLOOR COVER | S.F. | — | — | x | x |
| 909.0 | CARPET | S.Y. | — | — | x | x |
| 912.0 | PAINTING | S.F. | — | — | x | x |
| 913.3 | WALL PAPER | S.F. | — | — | x | x |
| 1106.3 | DISHWASHER | EA | — | x | — | — |
| 1106.6 | REFRIGERATOR | EA | — | x | — | — |
| 1106.7 | RANGE | EA | — | x | — | — |
| 1106.10 | RANGE HOOD | EA | — | x | — | — |
| 1106.11 | MEDICINE CABINET | EA | x | x | — | — |
| 1500.0 | PLUMBING | L.S | — | — | x | x |
| 1502.0 | H.V.A.C. | L.S | — | — | x | x |
| 1601.0 | ELECTRICAL | L.S | — | x | x | x |
| | TOTAL DIRECT COSTS | | x | x | x | |

FIGURE 56

## SUMMARY OF ESTIMATE

| | | | |
|---|---|---|---|
| BUILDING | LOCATION | PAGE OF | |
| ARCHITECT | OWNER | BID DATE/TIME | |
| CUBICAL CONTENTS | NO. OF STORIES | COST PER CUBIC FOOT | ESTIMATOR |
| FLOOR, AREA, SQUARE FEET | | COST PER SQUARE FOOT | CHECKER |

| Code | Description | Units | 2 Labor | 3 Material | 4 Sub | 5 Name |
|---|---|---|---|---|---|---|
| 101.1 | Supervision | Wk | x | ——— | ——— | ——— |
| 102.1 | Temp. Office | Mo | ——— | ——— | x | x |
| 102.3 | Temp. Toilets | Mo | ——— | ——— | x | x |
| 102.4 | Temp. Phone | Mo | ——— | ——— | x | x |
| 102.5 | Temp Utilities | Mo | ——— | ——— | x | x |
| 103.1 | Layout | T.O | x | x | ——— | ——— |
| 103.2 | Punch List | T.O | x | x | ——— | ——— |
| 103.13 | Hand Tools | T.O | ——— | x | ——— | ——— |
| | | | | | | |
| | | | | | | |
| | Direct Costs Forward | | x | x | x | |
| | Total Labor | | x | x | x | |
| | Total Material | | x ← | | | |
| | Total Sub | | x ← | | | |
| | Subtotal | | x | | | |
| | 4% use tax | | x | | | |
| | 20% Labor Burden | | x | | | |
| | Construction Interest | | x | (Prime +2% for 1/3 Construction Period) | | |
| | Permits | | x | | | |
| | Overhead & Profit | | x | | | |
| | Bond | | x | | | |
| | **Total Price** | | x | | | |

FIGURE 57

## CONCLUSIONS

The methods described in this manual are based on estimating the structure in the logical order of construction activities. By using this logical order, the estimator can start with the first activity and proceed through the entire job without omitting activities. The examples of activities discussed in this manual make up the items usually estimated by a general contractor with the summary sheet used to bring the items estimated by the contractor and the subcontractor together. To become efficient in this method, the estimator must be able to build the structure on paper piece by piece. He must be able to picture the structure in his mind, accounting for all the labor and material costs associated with the structure.

This book has covered the first of two courses on estimating. Certain essential basics of format, organization and estimating procedures have been presented. The reader is now ready to undertake the next estimating course which covers estimating a large commercial building with more complex plans and specifications.

# BIBLIOGRAPHY


Clough, Richard H., *Construction Contracting,* (John Wiley and Sons, New York, 1975).

Muller, Edward J., *Reading Architectural Working Drawings,* (Prentice-Hall Inc., New Jersey, 1971).

Park, William R., *Construction Bidding for Profit,* (John Wiley and Sons, New York, 1977).

Randall, Frank A., *Concrete Masonry Handbook,* (Portland Cement Association, Skokie, Illinois, 1976).

*Standard Building Code 1976 Edition,* (Southern Building Code Congress International, Birmingham, Alabama, 1976).

Walker, Frank R., *The Building Estimator's Reference Book,* 21st Ed., (Frank R. Walker Co., Chicago, Ill., 1982).

# APPENDIX - A
# SAMPLE COST CODING SYSTEM

## DIVISION #1 - GENERAL REQUIREMENTS

| | |
|---|---|
| 0101.0 | GENERAL CONDITIONS (A.I.A. Doc. A701 April 1970) |
| .1 | Superintendent |
| .2 | Permits & Fees |
| .3 | Taxes - Sales & Use Taxes |
| .4 | Taxes - Unemployment Compensation - State |
| .5 | Taxes - Unemployment Compensation - Federal |
| .6 | Taxes - F.I.C.A. - Employer's Share |
| .7 | Insurance - Contractors Liability (Worl. Comp. PL & PD) |
| .8 | Insurance - Hold Harmless |
| .9 | Insurance - Property (Fire & Extended Coverage) |
| .10 | Cleanup - Progress (CSI 01710) |

| | |
|---|---|
| 0102.0 | SPECIAL CONDITIONS (Temporary Facilities & Utilities) |
| .1 | Temporary Office (CSI 01590) |
| .2 | Temporary Storage & Protection (CSI 01630) |
| .3 | Temporary Sanitary Facilities (CSI 01516) |
| .4 | Temporary Telephone |
| .5 | Temporary Power & Light (CSI 01511) |
| .6 | Temporary Water (CSI 01515) |
| .7 | Temporary Heat (CSI 01513) |
| .9 | Temporary Roads and Ramps (CSI 01550) |

| | |
|---|---|
| 0103.0 | MISCELLANEOUS CONDITIONS |
| .1 | Field Engineering & Layout |
| .2 | Timekeepers & Accountants |
| .7 | Office Supplies & Equipment |
| .12 | Small Tools & Supplies |
| .13 | Hauling Equipment and Tools |
| .14 | Gas & Oil for Equipment |
| .17 | Scheduling - C.P.M., Pert., Network Analysis, etc. |
| .19 | Performance Bond - Subcontractors |
| .20 | Punch List & Final Out |

## DIVISION #2 - SITE WORK

| | | |
|---|---|---|
| 0201.0 | CLEARING OF SITE | |
| .1 | Structure Moving | L.S. |
| .2 | Clearing & Grubbing | ACRE |

| | | |
|---|---|---|
| 0202.0 | EARTH WORK | |
| .1 | Grading | |
| .11 | Machine Grading | S.Y. |
| .12 | Hand Grading | S.F. |
| .2 | Excavation | |
| .21 | Machine Excavation | C.Y. |
| .22 | Hand Excavation | C.Y. |
| .3 | Back fill | C.Y. |
| .4 | Fill | C.Y. |
| .5 | Compaction | C.Y. |
| .6 | Disposal | L.S. |
| .7 | Tests & Subsurface Exploration | L.S. |
| 0205.0 | SITE DRAINAGE | |
| .1 | Ditching | L.S. |
| .2 | Culverts | L.F. |
| .3 | Catch Basins & Man Holes | Ea. |
| .4 | Foundation Drainage | L.S. |
| .5 | Dewatering | L.S. |
| 0209.0 | PAVEMENTS & WALKS | |
| .1 | Paving | S.Y. |
| .2 | Curbs | L.F. |
| .3 | Walks | S.F. |
| 0209.0 | LANDSCAPING | |
| .1 | Seeding & Sodding | S.F. |
| .2 | Fertilizing | S.F. |
| .3 | Sprigging & Mulching | S.F. |
| .4 | Topsoil | C.Y. |
| .5 | Trees & Shrubs | L.S. |

## DIVISION #3 - CONCRETE

| | | |
|---|---|---|
| 0301.0 | CONCRETE CAST IN PLACE | |
| .1 | Concrete Placing | |
| .12 | Footings | C.Y. |
| .13 | Walls & Grade Beams | C.Y. |
| .14 | Columns & Pedestals | C.Y. |
| .15 | Beams | C.Y. |
| .16 | Slabs (Structural) | C.Y. |
| .17 | Stairs & Landing | C.Y. |
| .18 | Slabs on Ground | C.Y. |

| | | |
|---|---|---|
| .19 | Slabs over Decks or Lath | C.Y. |
| .20 | Toppings | C.Y. |
| .2 | Concrete Finishing (Including Screeds) | |
| .21 | Rough Screeding | S.F. |
| .22 | Trowel Finishing | S.F. |
| .23 | Brush or Broom Finishing | S.F. |
| .24 | Float Finishing | S.F. |
| .25 | Special Finishing (Rub, Patch, Retard, Sand Blast, Etc.) | |
| .26 | Curing | S.F. |

| | | |
|---|---|---|
| 0302.0 | FORMWORK | |
| .12 | Footings | S.F.C.A. |
| .13 | Walls & Grade Beams | S.F.C.A. |
| .14 | Columns & Pedestals | S.F.C.A. |
| .15 | Beams | S.F.C.A. |
| .16 | Slabs | S.F.C.A. |
| .17 | Stairs & Landing | S.F.C.A. |
| .18 | Slab on Ground | S.F.C.A. |
| .3 | Specialties | |
| .31 | Nails, Wire, Ties | L.S. |
| .32 | Anchors & Inserts | E.A. |
| .33 | Chamfers, Drips, Etc. | L.F. |
| .34 | Stair Nosings; Treads | L.F. |
| .35 | Water Stops | L.F. |
| .36 | Membrane & Waterproofing for Slabs | S.F. |

| | | |
|---|---|---|
| 0303.0 | REINFORCING STEEL | |
| 1. | Bars | Lb. |
| 2. | Wire Fabric | S.F. |
| 3. | Accessories (Chairs & Tie Wire) | L.S. |

## DIVISION #4 - MASONRY

| | | |
|---|---|---|
| 0401.0 | BRICK MASONRY | Ea. |
| 0402.0 | CONCRETE BLOCK & PRECAST LINTELS & SILLS | |
| 0406.0 | FLUE LINING | Ea. |
| 0407.0 | STONEWORK | Ea. |
| 0409.0 | MORTARS | Ea. |

| | | |
|---|---|---|
| 0410.0 | FILLED CONCRETE BLOCK | C.Y. |
| 0412.0 | ACCESSORIES & SPECIALTIES | |
| .1 | Anchors | Ea. |
| .2 | Ties | Ea. |
| .3 | Reinforcement | L.F. |
| .4 | Fire Place Accessories | |
| 0416.0 | EQUIPMENT & SCAFFOLDING | |

## DIVISION #5 - METALS

| | | |
|---|---|---|
| 0501.0 | STRUCTURAL METALS | |
| 1. | Beams & Columns | LB. |
| 2. | Channels | LB. |
| 3. | Angles | LB. |
| 4. | Structural T's | LB. |
| 5. | Structural Tubing | LB. |
| 5. | Plates & Bars | LB. |
| 0506.0 | METAL SIDINGS & ROOFINGS | |
| .1 | Steel | SQ. |
| .2 | Aluminum | SQ. |
| 0507.0 | LIGHT GAGE FRAMING | L.F. |
| 0508.0 | EXPANSION CONTROL | L.F. |

## DIVISION #6 - WOOD & PLASTICS

| | | |
|---|---|---|
| 0601.0 | ROUGH CARPENTRY | |
| .1 | Light Framing & Sheeting | |
| .11 | Studs & Plates | BF. |
| .12 | Joists & Headers | BF. |
| .13 | Bridging | BF. |
| .14 | Rafters | BF. |
| .15 | Stairs | BF. |
| .16 | Roof Sheatings | S.F. |
| .17 | Wall Sheating | S.F. |
| .18 | Sub Floor Sheating | S.F. |
| .19 | Roof Trusses | Ea. |
| .2 | Heavy Framing & Decking | |

| | | |
|---|---|---|
| .21 | Columns & Beams | BF. |
| .22 | Decking | BF. |
| .3 | Miscellaneous Carpentry | |
| .31 | Blocking & Bucks | BF. |
| .32 | Ground, Furring & Sleepers | L.F. |
| .33 | Cant Strips | BF. |
| .34 | Building Papers | S.F. |
| | | |
| 0602.0 | FINISH CARPENTRY | |
| .1 | Finishing Siding & Facing Materials | |
| .11 | Boards | BF. |
| .12 | Plywood | S.F. |
| .13 | Hardboard | S.F. |
| .14 | Cement Asbestos | S.F. |
| .15 | Shingles | S.F. |
| .16 | Facias | L.F. |
| .17 | Sofits | S.F. |
| .2 | Finish Walls & Facing Materials (Interior) | |
| .21 | Boards | S.F. |
| .22 | Hardboards or Pressed Wood | S.F. |
| .23 | Plywood Paneling | S.F. |
| .24 | Gypsum Board | S.F. |
| .25 | Vinyl Faced Paced Paneling | S.F. |
| | | |
| 0603.0 | ARCHITECTURAL WOOD WORK | |
| .1 | Custom Cabinet Work | L.S. |
| .2 | Counter Tops | L.S. |
| .3 | Custom Doors & Frames | Ea. |
| .4 | Moulding & Trim Facings | L.F. |
| .5 | Stairs | L.F. |
| .6 | Shelving | S.F. |
| .7 | Custom Paneling | S.F. |
| | | |
| 0607.0 | ROUGH HARDWARE | |

## DIVISION #7 - THERMAL & MOISTURE PROTECTION

| | | |
|---|---|---|
| 0701.0 | WATERPROOFING | |
| .1 | Membrane | S.F. |
| | | |
| 0702.0 | DAMPROOFING | |
| .1 | Bituminus | S.F. |
| .2 | Cementitious | S.F. |

| | | |
|---|---|---|
| .3 | Preformed Vapor Barriers | S.F. |
| .4 | Silicone | S.F. |
| | | |
| 0703.0 | BUILDING INSULATION | |
| .1 | Flexible (Fibrous Batts & Rolls) | S.F. |
| .2 | Rigid | S.F. |
| .3 | Loose | S.F. |
| .4 | Foamed | S.F. |
| .5 | Sprayed | S.F. |
| | | |
| 0704.0 | ROOF & DECK INSULATION | |
| | | |
| 0705.0 | MEMBRANE ROOFING | |
| .1 | Built Up | SQ. |
| .2 | Roll Roofing | SQ. |
| | | |
| 0707.0 | SHINGLE & TILE ROOFINGS | |
| .1 | Asphalt & Fibre Glass | S.F. |
| .2 | Asbestos Cement | S.F. |
| .3 | Wood | S.F. |
| .4 | Slate | S.F. |
| .5 | Tile | S.F. |
| | | |
| 0708.0 | SHEET METAL & FLASHING | |
| .1 | Down Spouts, Gutters & Fittings | L.F. |
| .2 | Facing, Gravel Stop & Trim | L.F. |
| .3 | Flashings | S.F. |
| .4 | Sheet Metal Roofing | S.F. |
| | | |
| 0709.0 | ROOF ACCESSORIES | |
| .1 | Skylights | Ea. |
| .2 | Sky Domes | Ea. |
| | | |
| 0710.0 | SEALANTS | |
| .1 | Caulking | L.F. |
| .2 | Joint Filler & Gaskets | L.F. |

## DIVISION #8 - DOOR WINDOWS & GLASS

| | | |
|---|---|---|
| 0801.0 | HOLLOW METAL DOORS & FRAMES | |
| .1 | Custom | Ea. |
| .2 | Stock | Ea. |

| | | |
|---|---|---|
| 0802.0 | NON FERROUS DOOR & FRAMES | |
| .1 | Aluminum | Ea. |
| .2 | Bronze | Ea. |
| .3 | Stainless Steel | Ea. |
| 0803.0 | WOOD & PLASTIC DOORS | |
| .1 | Flush Doors | Ea. |
| .2 | Panel Doors | Ea. |
| .3 | Louvered Doors | Ea. |
| 0804.0 | SPECIAL DOORS | |
| .1 | Closet Doors | Ea. |
| .2 | Glass - All | Ea. |
| .3 | Shower Doors | Ea. |
| .4 | Sliding or Patio Doors | |
| .41 | Metal (Aluminum) | Ea. |
| .42 | Wood (Inc. Insulating Glass) | Ea. |
| 0805.0 | METAL WINDOWS | |
| .1 | Projected (Casement) | Ea. |
| .2 | Sliding or Horizontal Rolling | Ea. |
| .3 | Double & Single Hung or Vertical Sliding | Ea. |
| .4 | Reversible (Aluminum) | Ea. |
| 0806.0 | WOOD & PLASTIC WINDOWS | |
| .1 | Basement Units | Ea. |
| .2 | Casement or Awning | Ea. |
| .3 | Double Hung | Ea. |
| .4 | Glider | Ea. |
| .5 | Picture View Window (with vent) | Ea. |
| .6 | Picture Window | Ea. |
| 0807.0 | SPECIAL WINDOWS | |
| .1 | Jalousie | Ea. |
| .2 | Pass Window | Ea. |
| 0808.0 | DOOR AND WINDOW ACCESSORIES | |
| .1 | Storm & Screens (Metal) | Ea. |
| 0810.0 | FINISH HARDWARE | |
| 0811.0 | WEATHER STRIPPING | Ea. |
| 0812.0 | GLASS & GLAZING | Ea. |

## DIVISION #9 - FINISHES

| | | |
|---|---|---|
| 0901.0 | LATH & PLASTER | |
| .1 | Lathing & Furring | |
| .11 | Metal Lath | S.Y. |
| .12 | Gypsum Lath | S.Y. |
| .14 | Metal Trim | L.F. |
| .2 | Plastering | |
| .21 | Gypsum | S.Y. |
| .22 | Accoustical | S.Y. |
| .23 | Vermiculate | S.Y. |
| .24 | Stucco or Cement Plaster | S.Y. |
| .25 | Thin Set | S.Y. |
| 0902.0 | GYPSUM DRYWALL | |
| .1 | Backer Board | S.F. |
| .2 | Finish Board | S.F. |
| .3 | Vinyl Coated Board | S.F. |
| .4 | Metal Stud Partition | S.F. |
| .6 | Metal Trim | L.F. |
| .7 | Dry Wall Finishing | S.F. |
| 0903.0 | TILE WORK | |
| .1 | Ceramic Tile | S.F. |
| 0906.0 | VENEER STONE & BRICK (FLOORS, WALLS, STOOLS, STAIRS, PARTITIONS) | |
| 0908.0 | RESILIENT FLOORING | |
| .1 | Linoleum | S.Y. |
| .2 | Vinyl | S.F. |
| .3 | Vinyl Asbestos | S.F. |
| 0909.0 | CARPETING | |
| 0910.0 | WOOD FLOORING | |
| .1 | Strip & Plank Flooring | S.F. |
| .2 | Parquet Flooring | S.F. |
| .3 | Finishing (Sanding, Staining, Sealing & Waxing) | S.F. |
| 0912.0 | PAINTING | |

| | | |
|---|---|---|
| 0913.0 | WALL COVERING | |
| .1 | Vinyl Plastic | S.F. |
| .2 | Plastic Wall Tiles | S.F. |
| .3 | Wall Paper | S.F. |
| .4 | Fabrics | S.F. |

## DIVISION #11 - RESIDENTIAL EQUIPMENT

| | | |
|---|---|---|
| 1106.0 | RESIDENTIAL EQUIPMENT | |
| .1 | Clothes Dryer | Ea. |
| .2 | Clothes Washer | Ea. |
| .3 | Dishwasher | Ea. |
| .4 | Compactor | Ea. |
| .5 | Freezer | Ea. |
| .6 | Refrigerator | Ea. |
| .7 | Range | Ea. |
| .8 | Oven | Ea. |
| .9 | Disposer | Ea. |
| .10 | Range Hoods | Ea. |
| .11 | Medicine Cabinets | Ea. |

## DIVISION #12 - FURNISHINGS

1202.0 BLINDS & SHADES

1204.0 DRAPERIES & CURTAINS

## DIVISION #13 - SPECIAL CONSTRUCTION

1303.0 CHIMNEYS - SPECIAL
- .1 Job Construction
- .2 Prefabricated

1318.0 SWIMMING POOLS

## DIVISION #15 - MECHANICAL

1501.0 PLUMBING
1502.0 HEATING AND AIR CONDITIONING

# DIVISION #16 - ELECTRICAL

1601.0 ELECTRICAL

## ABBREVIATIONS AND SYMBOLS

| ABBREVIATION | EXPLANATION |
|---|---|
| A/E | Architect/Engineer |
| BF | Board Feet |
| BM | Board Measure |
| Conc | Concrete |
| CF or Cu Ft | Cubic Feet |
| CY or Cu Yd | Cubic Yards |
| FBM | Feet Board Measure |
| G. C. | General Contractor |
| Hrs | Hours |
| HP | Horsepower |
| L | Labor |
| L. F. | Linear Feet |
| L. S. | Lump Sum |
| M | Material |
| O.C. | On Center |
| O.H. | Overhead |
| Pcs | Pieces |
| psf | Pound Per Square Foot |
| Ref | Reference |
| SHT. | Sheet |
| Spec. Sect. | Specifications Section |
| SF or Sq Ft | Square Feet |
| SFCA | Square Feet of Contact Area |
| S or Sub. | Subcontractor |
| T.O. | Take Off |
| MFBM | Thousand Feet Board Measure |
| Tns. | Tons |